Elias Colbert

Astronomy without a telescope

Elias Colbert

Astronomy without a telescope

ISBN/EAN: 9783744737388

Printed in Europe, USA, Canada, Australia, Japan

Cover: Foto ©berggeist007 / pixelio.de

More available books at **www.hansebooks.com**

ASTRONOMY

WITHOUT A TELESCOPE:

BEING

A GUIDE-BOOK TO THE VISIBLE HEAVENS,

WITH ALL NECESSARY

MAPS AND ILLUSTRATIONS.

DESIGNED FOR THE USE OF SCHOOLS.

BY

E. COLBERT.

CHICAGO:
GEORGE & C. W. SHERWOOD.
1869.

Entered according to Act of Congress, in the year 1869,
By GEO. & C. W. SHERWOOD,
In the Clerk's Office of the District Court of the United States, for the Northern District of Illinois.

CHURCH, GOODMAN AND DONNELLEY, PRINTERS, CHICAGO.   BOND AND CHANDLER, ENGRAVERS.

PREFACE.

There are four ways of regarding a subject — with the eye of sense, of reason, of fancy, and of faith. These severally constitute direct, inductive, dreamy, and substituted vision. The first is in itself interesting, yet not complete, or always reliable; but it is especially valuable as being the basis of all that is true in the second, consistent in the third, and credible in the last.

While in other departments of scientific tuition, the eye of sense is first appealed to, the would-be student of astronomy has generally been first taught to use some one of the other three; dissertations on the profundities of space and the acme of complicated motion, views of elaborately engraved monstrosities, with long chapters from the heathen mythology, and obscure descriptions of the invisible; all these, instead of a simple delineation of the heavens, as visible to the naked eye. It is thus that astronomy, naturally pleasing, and normally useful, is a sealed book, except to the select few — a general ignorance due altogether to the lack of aid to the study.

The following work is intended as an introduction to an acquaintance with the visible heavens; to educate the eye into an ability to recognize the most prominent stellar groups. The maps are small and simple; they present, in usable form, all the star groupings ordinarily visible to the naked eye in the United States, showing them, with little distortion, in the same relative positions as they occupy in the firmament, and without cumbering them up with painted figures — the form of the constellation is given only in faint outline. The greater portion of the maps are projected on the ecliptic as a base line, enabling the student to refer each star directly to the earth's path, and to trace out readily the course of a planetary body. The maps are so connected that a transition from one to another is easy, and the necessary explanatory text lies in, or near, the same opening of the book as the matter explained, while conciseness is aimed at throughout.

The solar system is treated visually and deductively. No attempt has been made to describe positions or appearances inappreciable to the naked eye, except so far as they are necessary to a proper comprehension of that which can be seen by the unassisted vision. The phenomena of the heavenly motions are explained, and concise aids are furnished for determining, almost at a glance, the positions of the principal members of the solar family among the fixed stars, and with reference to the meridian and horizon, at any time during the last thirty years of the nineteenth century. The discussion of eclipses, and the elucidation of the means whereby we can calculate the absolute distances and bulks, and comparative weights of some of the heavenly bodies, are within the range of the simpler mathematics; the only points involved, not included in the ordinary arithmetic, being the ratio of areas to squares, and contents to cubes, of like dimensions of similar figures, the composition of momenta, the analogies of plane and spherical right-angled triangles, and the addition of logarithms to find the product of their corresponding numbers. The elements of distance and bulk have been recalculated to the latest determinations of the solar parallax.

The text is susceptible of verbal amplification, at the discretion of the teacher. The foot-notes are intended to serve, at once, as a table of contents, and as the groundwork of questions on the text. When they are used interrogatively, the pupil should be required to give a full explanation or definition of the subject referred to in each note. It is recommended that the student transfer the principal star groups to slate or paper, without outlines or letters.

CONTENTS.

EARTH'S MOTION; on her own axis, and around the Sun; Phenomena; Circles of the Sphere; Angular Measures. — 5–9
FIXED STARS. Distances and Names; Sun's Annual Journey along the Ecliptic; Table. — 10–12
MAP No. I. Stars near the North Pole; the Dipper; Chair, Ursa Minor; Cepheus; Draco. — 13–16
MAP No. II. Stars near the Vernal Equinox; Square of Pegasus; Perseus; Andromeda; Lacerta; Pisces; Aries; Musca; Trianguli. — 17–19
MAP No. III. The Zodiac; Cetus; Phoenix; Eridanus. — 20–21
MAP No. IV. Three to six hours in Right Ascension; Perseus; Taurus; The Pleiades; Auriga; Orion; Hydrus; Reticulum Rhomboidalis; Doradus. — 22–24
MAP No. V. Gemini; Telescopium Herschelium; Cancer; Canis Minor; Canis Major; Lepus; Columna Noachi; Lynx; Camelopardalus. — 25–27
MAP No. VI. Ursa Major; Canes Venatici; Leo; Leo Minor. — 28–29
MAP No. VII. Virgo; Coma Berenices; Crater; Corvus; Hydra. — 30–31
MAP No. VIII. Libra; Scorpio; Ara; Triangulum Australis; Ophiucus et Serpens; Norma Euclidis; Circinus. — 32–34
MAP No. IX. Taurus Poniatowski; Aquila et Antinous; Delphinus; Sagittarius; Corona Australis; Microscopium; Grus; Indus et Pavo; Telescopium. — 35–36
MAP No. X. Capricornus; Aquarius; Piscis Australis; Pegasus; Equuleus. — 37–38
MAP No. XI. Boötes; Corona Borealis; Hercules; Lyra. — 39–40
MAP No. XII. Argo; Robur Caroli; Stars near the South Pole; the Milky Way. — 41–42
MAP No. XIII. Cygnus; Centaurus et Crux; Lupus. — 43–44
TABLE OF FIXED STARS. Magnitudes; Names; Right Ascensions, North Polar Distances, Longitudes, and Latitudes on January 1st, 1875; Annual Variations. — 45–53
CULMINATING, RISING AND SETTING. Ascensional Differences; Refraction; To Find Right Ascension and Declination from Longitude and Latitude. — 54–59
MAP No. XIV. Principal Stars Visible in North Latitude 40°; Positions with reference to the Meridian and Horizon. — 60–61
THE SOLAR SYSTEM. Jupiter, Venus, Mercury, Mars, Saturn, Uranus, Neptune; Phenomena of Motion; Heliocentric and Geocentric Positions; Tables of Places from 1870 to 1900. — 62–70
THE MOON. Lunation; Parallax; Eclipses; Measures of Time; Chronology. — 71–78
DISTANCE AND BULK. Measuring the Earth and Moon; Sun's Parallax; Earth's Orbit; The Sun; Laws of Motion; Mutual Attraction and its Results; Precession; The Tides. — 79–88
ELEMENTS OF THE SOLAR SYSTEM. Distances, Magnitudes, Bulks, etc., of the Planets and Luminaries. — 89–90
OTHER MEMBERS OF THE FAMILY. Planetoids; Rings, Satellites, Comets; Aerolites. — 91–95
THE FIXED STARS. Parallax; Distances, Numbers, Magnitudes; Light; Individual Motion; Common Origin. — 96–98
ABSOLUTE MOTION OF THE SOLAR SYSTEM. Sun's Orbit; Real Paths of the Earth and Moon. — 99–100
INDEX. — 101–104

INTRODUCTORY.

1. The study of Astronomy includes two classes of objects — the stationary, and the wandering.

2. The first class consists of the Fixed Stars; so called, not because they are absolutely immovable, but for the reason that they change their relative positions very slowly, a thousand years making but little difference in their apparent locations with regard to each other.

3. The second class comprises the Sun, Moon, Planetary Bodies, Asteroids, Comets, etc.; they change their apparent places among the fixed stars with varying degrees of rapidity. Their paths in the heavens can only be traced out by one who has an ocular acquaintance with the leading members of the first named class.

4. The fixed stars appear to be set in the surface of a hollow sphere. If we watch them attentively for an hour or two, on a bright night, we notice a connected movement. Those in the Eastern part of the heavens are slowly rising, those in the South are passing towards the West, and those in the West are sinking towards the horizon. If, standing any where in the Northern States, we face the North, and look up about half-way between the point overhead and the horizon, we may select any star, and, following its movements, see that it travels around but a small circle; and there is one dull red star, shining out timidly in the midst of comparative blankness, whose circle of apparent travel is so small, that the aid of an instrument is necessary to enable us to detect a movement. This is called the POLE STAR.

5. This apparent movement of the stars is caused by the motion of the earth on its axis. The earth is a large globe, nearly eight thousand miles in diameter (the greatest distance through it is 41,847,200 feet). We live on its exterior, the upright position at any place being perpendicular to the surface at that spot; the centre of the globe is, every where, the point to which all bodies tend to fall, in obedience to the law of attraction of gravitation. The earth turns round once in twenty-four hours, in just the same way that a school globe is rotated on its axis; except that the earth moves in space, with no frame work to support it.

6. Take a round ball, of yarn (or an orange), and run a piece of wire through its centre, making it pass entirely through the ball. Hold one end of the wire firmly in the hand, pointing the other end in the direction of the North, but sloping upwards, about midway between the level and upright positions; turn the ball slowly round on the wire. If a small insect were gummed fast to the surface of the ball, at a little distance from the northern end of the wire, he would preserve his position with regard to the centre of the ball, but, at each rotation, every visible object would appear to move round in a circle, the centre of which would be in the wire.

7. Enlarge the ball to a globe of nearly eight thousand miles in diameter, replace the insect by a human being, held to the surface by the attraction of gravitation, let the globe rotate in space, instead of being sustained on a wire, and let the stars be the surrounding objects; we have the earth in motion. The two points on the surface, corresponding to those cut by the wire, are called THE POLES. A line passing from one pole to the other would be the earth's axis; a circle traced around the surface, equally distant from the two poles, and marking off the globe into two halves, each of which would have one pole in the centre of its curved surface, would coincide with the equator. A circle passing around the globe, through both poles, and dividing the equator into two equal semi-circles, would be a circle of the meridian.

8. A circle of the meridian passing through any point on the earth's surface, is the meridian of that place; and when a straight line, joining the centres of the earth and a star, would cut this circle at any point in its circumference, the star is said to be on the meridian of that place.

9. If the line of the earth's axis were extended to the fixed stars, it would mark two opposite points which never change their position; these are the Celestial Poles. The one adjacent to the pole star (Sec. 4) is called the North Pole; the point opposite to it is the South Pole. Similarly, a line joining any two points in the Equator (Sec. 7), if extended to the stars, would mark out among them, in the course of one rotation, a circle called the Equinoctial — every where equally distant from each of the celestial poles. A circle passing through both poles would divide the Equinoctial into two equal semi-circles; it would be a circle of the meridian.

10. The expressions "would be," and "were," are employed in the above descriptions, for the reason that the polar points and the circles mentioned, are not actually marked, either on the earth or in the heavens; in both cases, the points through which the lines referred to would pass, are determined by measuring from the fixed stars. Thus, we find the place of the North Pole by accounting it to be at a certain distance, in a certain direction, from the Pole Star.

11. There is a second general movement observable among the stars. If we watch a group situated near the Equinoctial, noting its position by sighting it along a wall at, say, 9 o'clock this evening, a week hence we shall find it arriving at the same apparent place, at a little after half past eight o'clock; and a month hence it will be about 7 o'clock instead of 9. Six months hence, the stars now overhead at 9 o'clock in the evening, will be overhead at 9 o'clock in the morning; and not till next year, at this time, will the same place be occupied at 9 o'clock in the evening, as now. The fixed stars gain nearly four minutes (3 min. 56 sec.) daily, or two hours per month, or twenty-four hours in a year. On what do they gain?

12. On the Sun; our time reckoning is measured by his apparent motion in the heavens. The earth, while performing a daily rotation on her own axis, makes, each year, a revolution around the Sun. If we could stand on the solar surface, and view the earth, we should see her slowly moving forward, apparently among the fixed stars, making an entire circuit of the firmament in a little more than 365 days (365 d. 5 h. 48 m. 49 sec.). To us, standing on the earth, it is the Sun that appears to make the journey. When he is on the meridian (Sec. 8), we say it is noon, and the time between any two consecutive appearances of the Sun on the meridian is one day, which we artificially divide into twenty-four hours. But in that interval of time, the Sun has apparently passed over nearly one part in 365 of his annual journey, changing his place among the fixed stars by that amount, and a star near his apparent path will arrive on the meridian, on the second day, as much sooner as is given by the following proportion:

As the time of the annual circuit is to one day, so is the time of one diurnal rotation to the time gained by a star; or,

As 365 d. 6 h. is to 1 day, so is 24 hours to 3 min. 56 sec., nearly.

THE EARTH'S MOTION.

13. The Sun's pathway among the stars is called the ECLIPTIC. This circle does not coincide with the Equinoctial, but makes an angle with it of a little more than one quarter of a right angle. We may illustrate the relations of the two circles by recurring to the ball and wire (Sec. 6). Let a lamp, placed nearly in the centre of a large round table, represent the Sun, and the circumference of the table the path of the earth. Hold the wire pointing northward, as before, but let it slope less from the perpendicular — leaning only about one fourth of the angle towards the horizon. Then move the hand slowly around the edge of the table, always keeping the wire at the same inclination towards the North, and turning the ball round on the wire at the same time. The slope of the wire, as referred to the flat surface of the table, will represent the position of the earth's axis with regard to a plane surface passing through every point in the Ecliptic circle.

14. The Ecliptic and Equinoctial intersect each other in two opposite points, which are passed over by the Sun, respectively, about the 20th of March and the 23rd of September in each year. These are called the Equinoxes, because when the Sun is passing those points, the days are about equal in length to the nights, all over the globe.

15. The following diagram shows the position of the earth's axis with regard to the plane of the Ecliptic: The central figure represents the globe, with the North Pole (at A) pointing in an oblique position. The curve B C D represents one half of the equator; the globe rotates in the direction B to C, C to D, the journey from B to C being made in about six hours, that from C to D in six hours, and the entire rotation, from B to B again, in a little less than twenty-four hours. The surrounding circle represents the Ecliptic, the page on which it is printed corresponds to the plane of the Ecliptic, and passes through the centre of the

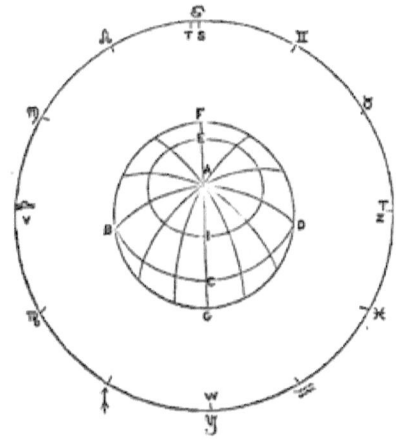

globe, cutting the Equator in the points B and D. If the earth simply rotated on her own axis, then, to a spectator at E, it would be noon when the Sun was at S, the luminary then being in a right line with the earth's centre, and F, which is a point in the meridian F E A I C G; six hours afterwards the rotation of the earth would carry that meridian round to the position of the curved line G A D, and the Sun would then be near the western horizon; six hours still later, the meridian would be carried half way round, and the spectator would be at I, it being then midnight; in twelve hours more the earth would have completed an entire rotation, bringing the spectator again to C, with the Sun at S on the meridian.

16. But owing to the annual revolution, it will still lack nearly four minutes of noon, because during the time of one rotation on her axis, the earth's motion around the Sun will have caused him to appear to have moved forward in the Ecliptic to T. A fixed star at S would return to the meridian at equal intervals of 23 h. 56 m. 4 sec., and make 366¼ revolutions, while the Sun made 365¼ apparent daily circuits, just as the minute hand of a clock passes round the dial thirteen times in passing the hour hand twelve times. The intervals between the successive returns of the Sun to the meridian of any place average twenty-four hours; but these intervals are not all exactly equal. The Sun is not in the exact centre of the earth's orbit, and as she moves more rapidly when on that side of her orbit nearest to him, than when at her farthest distance, the Sun's apparent daily motions in the Ecliptic are unequal, and that body is sometimes on the meridian a little before, and at other parts of the year a little after, the time indicated as noon by a clock which measures equal days of 24 hours each. This difference is called the Equation of time, and its direction is indicated by the expressions "Clock fast," or "Clock slow." The difference between Mean (clock) noon, and Solar (Sun) noon, is given in most almanacs, for each day of the year.

17. The sun is seen in the position S, about the 21st of June; the North Pole, A, is then turned partially towards him, and to a spectator at any part of the surface between the pole and the equator, B C D, he is longer above the horizon than below. It is midsummer; the stars at W are seen on the meridian at midnight. Three months afterwards — September 23rd — the Sun will have progressed to the position V; he is then on the Equinoctial, in the plane of the Equator, B C D, and the days and nights are of equal length. The stars at Z are now seen on the meridian at midnight. In three months more — December 22nd — the Sun is seen at W; the north pole is then partially turned away from him, and the nights in the northern hemisphere are longer than the days; it is midwinter, but those living on that half of the earth's surface which is nearest to the south pole, have the long days and the short nights, with the warm weather of the summer season. The stars at S, which were obscured by the Sun's presence in June, are now visible on the meridian at midnight. Still another three months brings the Sun to Z — March 20th — where he is again on the Equinoctial, and the days and nights are once more equal. This is the Vernal, or Spring Equinox.

18. The two great circles, the Ecliptic and Equinoctial, thus intersect each other at the points Z and V. Astronomers have agreed to take the first — the Vernal Equinox — as the one from which the distances of the heavenly bodies shall be measured, and to which they shall be referred for comparison of their mutual positions. The measurements are made on both circles, proceeding in the direction Z S V W Z, and sideways from them towards the poles.

19. The distance from the Vernal Equinox, measured on the Equinoctial, is called Right Ascension.

20. The distance sideways from the Equinoctial is called Declination, and is called "North," or "South," according as it is measured towards the North or South Pole. Sometimes (see Sec. 136), instead of the Declination, it is preferred to use the distance from the North Pole (abbreviated to "Nor. Pol. Dis."). Both are measured on a circle of the meridian passing through the apparent centre of the star.

21. Distance from the Vernal Equinox, measured on the Ecliptic, is called Longitude. The distance sideways from the Ecliptic is called Latitude, and is North or South as in the case of Declination.

22. These dimensions are not taken in feet and inches, or miles. We do not attempt, in this connection, to measure actual distances, but to find the magnitude of the angle formed by straight lines, from stars or points, meeting at the eye of the observer. For this purpose the circle is supposed to be divided into 360 equal parts, called Degrees, and each of these degrees into 60 equal parts called Minutes, and each minute into 60 equal seconds. A quarter of the circumference of the circle, or 90 degrees, is the measure of a right angle. The angle included between the planes of the Ecliptic and Equinoctial is 23 degrees, 27 minutes, and 20 seconds; thus expressed: 23° 27' 20".

23. The Ecliptic is further divided into twelve equal parts, called Signs, each containing 30°. The following are their names in Latin, with the characters used to represent them. The signs are shown in their order, around the preceding diagram (Sec. 15), the character being placed at the beginning of the space denoted by it:

♈ Aries.   ♌ Leo.   ♐ Sagittarius.
♉ Taurus. ♍ Virgo. ♑ Capricornus.
♊ Gemini. ♎ Libra. ♒ Aquarius.
♋ Cancer. ♏ Scorpio. ♓ Pisces.

24. These signs must not be confounded with twelve star groups, bearing the same names, to be mentioned subsequently. The spaces were thus named by the early astronomers, many years before the Christian Era, to signify the character of the season experienced while the Sun was apparently passing through them; and the stars then within those limits were called by the same names.

But since that time the stars have all gradually moved forward, so that those then in ♈ are now in ♉, those then in ♉ are now in ♊, etc. Both Signs and star groups have retained the names then given, the group called "Aries" being now in the space called "Taurus." The Longitude of a star is often counted on the Ecliptic from 0° to 360°, but generally reckoned as so many degrees in a certain sign. Thus the Longitude 168° 3' 9" is expressed as ♍ 18° 3' 9"; that is, 150° for five whole signs, and the overplus counted as from the beginning of ♍, the sixth sign.

25. This third connected movement of the stars is much slower than either of the two already spoken of (Secs. 4 and 11). One revolution occupies many centuries, and the movement requires many years of patient watching to measure it; the amount of change is 50¼ seconds annually, or 1° in about 72 years. It results from the fact that the earth's axis, which we have hitherto considered as always pointing to the same part of the heavens, has a slow, swinging motion, describing a circle in the heavens, of a diameter equal to twice the angle formed by the Ecliptic and Equinoctial, in about 25,750 years. This causes the points of intersection of these two circles to recede along the Ecliptic, making one round in that time, and causing the stars to appear to move forward with respect to the Equinoctial points, from which their Longitudes and Right Ascensions are measured. This motion is called the Precession of the Equinoxes.

DEFINE AND EXPLAIN (the figures refer to the sections): 2. Fixed stars. 3. Moving bodies. 4. Diurnal motion; Pole star. 5. Earth's diameter, time of rotation; centre of gravitation. 6. Earth's motion in miniature. 7. Motion in space; axis; equator; circle of meridian. 8. Meridian of any place. 9. Celestial poles; equinoctial. 10. Imaginary lines. 11. Annual revolution; daily progress of stars. 12. Measure of time; year; earth journey; noon; divisions of day. 13. Ecliptic, angle with equinoctial; illustration of annual motion. 14. Equinoxes. 15. Simple rotation. 16. Compound and unequal motion; equation of time. 17. Position of Sun each three months; summer and winter. 18. Intersections of ecliptic and equinoctial; origin of measures. 19. Right Ascension. 20. Declination. 21. Longitude; latitude. 22. Angular measure, degrees in circle; subdivisions. 23. Signs; characters and names. 24. A necessary distinction; two modes of expressing longitude. 25. Precession of equinoxes; rate, effect.

26. Distances in Right Ascension are also reckoned in degrees and minutes, but it is generally more convenient to count them in time, the Right Ascension of a star being equal to the number of hours and minutes which elapse between the Meridian passage of the Vernal Equinox and that of the star. The clock used must, however, be one showing sidereal time; that is, marking the lapse of 24 hours in the same time that the ordinary clock notes 23 h. 56 m. 4 s., or between the two consecutive times of a fixed star passing the Meridian (See. 12). Twenty-four hours thus measures the passage of 360°; one hour, therefore, is equivalent to 15°, and 1° is equal in value to four minutes of time. The minutes and seconds of time are marked "m," and "s," to distinguish them from minutes and seconds of space (See. 22).

27. The Ecliptic is not only the line of the Earth's annual revolution, but it is also remarkable as being the middle of a narrow band, or zone, of about 16° in breadth, within which all the principal members of the solar system, and many of the smaller ones, are always found when seen from any point on the Earth's surface. It is hence important to be able to trace out its course among the fixed stars, the planets Jupiter and Saturn being always close to the Ecliptic, and the Moon, Venus, and Mars, never more than about 8° distant. For this reason most of the following maps are projected with the Ecliptic as the principal line, and a straight one; the Equinoctial, when within the limits of the map, running obliquely. The maps are constructed on a scale of 10°, on the Ecliptic, to the inch.

28. The fixed stars are scattered irregularly over the heavens, in uncounted numbers; those ordinarily visible to the naked eye are not numerous. The most prominent are mapped out in this book, with the position of an irregular band of minute stars and star clouds, running around the heavens, and called the Galaxy, or Milky Way.

29. The early observers of the stars divided them into groups, often fanciful, but still useful; those groupings are recognized, with but few changes, by astronomers of the present day. The shapes chosen were generally those of natural and common objects. We may not be able to discern lions, bears, goats, fishes, dogs, or men, among the stars, but for that we are not responsible; it is a very convenient arrangement, and one not yet improved upon.

30. Each of these groups is called a Constellation. The stars comprised in each are of various degrees of brightness — the greater prominence being probably due in some instances to relative nearness, in others to greater actual size. The largest in appearance are said to be of the first magnitude, those next in brightness are classed as of the second; those so faint as to be but just discernible with the naked eye under the most favorable circumstances, are enumerated as of the sixth magnitude; stars seen only by the aid of a telescope, are classed from the seventh to the sixteenth magnitudes, inclusive; and clusters of telescopic stars, or of matter which the telescope has not yet resolved into stars, are called Nebulæ. On the following maps are shown the relative positions of all stars from the first to the fifth magnitudes inclusive, within about 25° of the Ecliptic, and all stars of greater magnitude than the fifth, outside of those limits, with a few smaller ones in special cases, and a few of the more remarkable nebulæ. The apparent relative sizes are distinguished as follows:

Stars of the first magnitude, with eight rays.
Second mag., seven rays. Third mag., six rays.
Fourth mag., five rays.   Fifth mag., four rays.
Nebulous clusters, by a circle, enclosing a dot.

DISTANCES AND NAMES.

31. The Stars are individualized by giving proper names to the most prominent, and also by naming all with the letters of the Greek alphabet, usually giving the first letter to the largest star in the constellation, the next letter to the second star in apparent magnitude, and so to Omega, after which the Roman letters, and sometimes Arabic numerals, are employed. The name of the constellation is put in the genitive case. Thus the largest star in the constellation, Leo — the Lion — is called " Regulus," and enumerated as α (Alpha) Leonis. The following is the Greek alphabet.

NAME.		NAME.		NAME.	
α	Alpha.	ι	Iota.	ρ	Rho.
β	Beta.	κ	Kappa.	σ	Sigma.
γ	Gamma.	λ	Lambda.	τ	Tau.
δ	Delta.	μ	Mu.	υ	Upsilon.
ε	Epsilon.	ν	Nu.	φ	Phi.
ζ	Zeta.	ξ	Xi.	χ	Chi.
η	Eta.	ο	Omicron.	ψ	Psi.
θ	Theta.	π	Pi.	ω	Omega.

In a few cases one letter is applied to two or more stars in the same constellation, and followed by a small numeral; as $α^1$, and $α^2$. The stars so marked belong to a subordinate group, generally with an independent motion of its own, its members revolving around a common centre.

32. The point in the heavens directly overhead is called the Zenith. The direction "Northward," in the heavens, is always towards the North Pole, not to the northern point of the horizon. "Eastward" is always in the direction of *increased* Right Ascension, as from 11 hours towards 12 hours, and "Westward" is in the direction of *decreased* Right Ascension; the latter fact should be especially borne in mind in the case of stars between the Pole and the horizon, as then the direction is exactly the reverse of that understood by the use of the same words in reference to points on the Earth's surface.

33. The following table shows the Sun's Longitude, both as counted direct from the Vernal Equinox, and as reckoned by signs, with his Right Ascension in time, and the Equation of time, for the noon of the 1st and 15th days of the twelve months next preceding February 29th in Leap Year. The Longitude and Right Ascension for intermediate days may easily be found by proportion. The table is nearly correct for any year in the last half of the present century, and the Sun's place may be found very nearly by adding 45' to the Longitude, and 3 min. to the Right Ascension, for any day in the first twelve months after Leap Year's day; 30' and 2 min. for the second twelve months; and 15' for Longitude, and 1 min. for Right Ascension, for the third twelve months after February 29th, to the quantities given in the table.

34. The Right Ascension of the Sun on any day is equal to the time that the Sun passes the Meridian before the Vernal Equinox (Sec. 26). The Equation of Time (Sec. 16) is the difference between the clock and the Sun. If, therefore, we apply the Equation as given in the last column of the table, adding it to, or subtracting it from, the Sun's Right Ascension, as the sign is + (plus), or − (minus), we shall know the number of hours and minutes that the Vernal Equinox passes the meridian above the Earth, before Clock, or Mean, Noon. This corrected Right Ascension is called Sidereal Time. For all ordinary purposes of observation the Equation of time may be neglected.

35. From the Right Ascension of a Fixed Star (adding 24 hours, if necessary) subtract the Sidereal Time for the day required; the remainder is the time past noon when the Star will be on the Meridian above the Earth. To the Sidereal time (or the Sun's Right Ascension if accuracy be not required) add the time past noon; the sum will give the Right Ascension on the Meridian at that time.

ASTRONOMY.

Day.	Sun's Longitude.			Right Ascension.	Equation of Time.
March 1	340° 47'	≈	♓ 10° 47'	22h. 49m. 5s.	— 12m. 36s.
" 15	355° 47'	≈	♓ 25° 47'	23h. 44m. 31s.	— 8m. 48s.
April 1	11° 37'	=	♈ 11° 37'	0h. 42m. 41s.	— 4m. 15s.
" 15	25° 21'	=	♈ 25° 21'	1h. 34m. 1s.	— 0m. 1s.
May 1	40° 55'	=	♉ 10° 55'	2h. 33m. 56s.	+ 2m. 59s.
" 15	54° 27'	=	♉ 24° 27'	3h. 28m. 20s.	+ 3m. 51s.
June 1	70° 46'	=	♊ 10° 46'	4h. 36m. 33s.	+ 2m. 30s.
" 15	84° 10'	=	♊ 24° 10'	5h. 34m. 34s.	— 0m. 7s.
July 1	99° 25'	=	♋ 9° 25'	6h. 40m. 59s.	— 3m. 27s.
" 15	112° 46'	=	♋ 22° 46'	7h. 38m. 20s.	— 5m. 36s.
August 1	129° 1'	=	♌ 9° 1'	8h. 45m. 40s.	— 6m. 2s.
" 15	142° 27'	=	♌ 22° 27'	9h. 39m. 11s.	— 4m. 14s.
September 1	158° 51'	=	♍ 8° 51'	10h. 41m. 51s.	+ 0m. 9s.
" 15	172° 28'	=	♍ 22° 28'	11h. 32m. 20s.	+ 4m. 53s.
October 1	188° 9'	=	♎ 8° 9'	12h. 20m. 56s.	+ 10m. 21s.
" 15	201° 59'	=	♎ 21° 59'	13h. 21m. 16s.	+ 14m. 10s.
November 1	218° 56'	=	♏ 8° 56'	14h. 26m. 10s.	+ 16m. 18s.
" 15	233° 1'	=	♏ 23° 1'	15h. 22m. 27s.	+ 15m. 12s.
December 1	249° 12'	=	♐ 9° 12'	16h. 30m. 1s.	+ 10m. 43s.
" 15	263° 25'	=	♐ 23° 25'	17h. 31m. 18s.	+ 4m. 34s.
January 1	280° 45'	=	♑ 10° 45'	18h. 46m. 46s.	— 3m. 48s.
" 15	295° 1'	=	♑ 25° 1'	19h. 47m. 51s.	— 9m. 41s.
February 1	312° 18'	=	♒ 12° 18'	20h. 59m. 4s.	— 13m. 58s.
" 15	326° 28'	=	♒ 26° 28'	21h. 54m. 49s.	— 14m. 24s.
" 29	340° 33'	=	♓ 10° 33'	22h. 48m. 13s.	— 12m. 37s.

DEFINE AND EXPLAIN (the figures refer to the sections):

26. Right Ascension in time. 27. The Zodiac; its width and medial line; what it includes; projection of maps. 28. Distribution of fixed stars. 29. Constellations. 30. Classification of magnitudes. 31. Naming the stars; Greek alphabet; other marks. 32. Zenith; East and West in the heavens. 33. Sun's Longitude and Right Ascension through the year; equation of time. 34. Clock noon; Sidereal time. 35. When a star is on the meridian.

MAP NO. I.

36. We commence the work of tracing out the relative locations of the fixed stars, with the group called THE DIPPER — the most prominent part of a constellation known as Ursa Major. It is a well known and easily recognized assemblage of stars; is always above the horizon of the Northern, and of the northern tier of the Southern, States; lies directly across the Equinoctial Colure (a meridian circle passing through the Equinoxes, hence sometimes called the first celestial meridian), and forms an easy guide to the position of the Pole Star. The Dipper is shown in the lower right hand corner of the map. It is visible in this position, on moderately clear nights, at the time the Vernal Equinox coincides with the meridian above the earth, as shown in the preceding table. Thus, on December 1st, the Sidereal time is 16 h. 30 m. + 10¾ m. = 16 h. 40¾ m.; and at that time before noon, or at 7 h. 19½ m. in the evening, the Dipper occupies the position shown on the map; it is visible in nearly the same position for an hour before and after that time.

37. From this position the Dipper slowly rises towards the right, in an oblique direction, and six hours afterwards is in the North-eastern part of the heavens, and 30° to 40° above the horizon, with the handle pointing downward. Turn the map so that the Dipper will be in the upper right hand corner, and it will show the appearance then; for example, at 1 h. 10 m. in the morning of December 2nd. From this position the group still rises, till six hours later — as at 7 h. 19 m. in the morning of December 2nd — it is on the meridian, a few degrees north of the Zenith, with the handle towards the right; as shown by turning the map upside down, bringing the Dipper to the upper left hand corner. Six hours still later the Dipper occupies the position shown by turning the map so as to bring the group into the lower left hand corner, the handle pointing upwards, in the North-western quarter of the heavens. The cause of these changes, both in position and time, has been explained in Sections 5 and 12.

38. The seven prominent stars in the Dipper are lettered from a to η inclusive (Sec. 31), beginning at the top of the side furthest from the handle — the Western side (Refer to Sec. 32). Each has also a proper name. They are: a, Dubhe; β, Merak; γ, Pheeda; δ, Megrez; ϵ, Alioth; ζ, Mizar; η, Alkaid, called also Benetnasch. Dubhe and Merak are known as the POINTERS, because, in whatever part of its daily circuit the Dipper may be, they always point towards the North Pole. On the map this point is at the intersection of the two right lines, near the middle of the right hand margin. Near it is a dull red star of the 2nd magnitude (Sec. 30) marked a, and named Alruccabah. (Sec. 4.) It is so near the Pole that its place is often accepted as that of the Pole itself. It is 27° 31′ from Dubhe, and a little outside of a line from Merak through Dubhe. The distance may be gauged very nearly by remembering that it is about one tenth greater than that between Dubhe and Alkaid in the extremity of the handle. The distance from Dubhe to Alkaid is 25° of a great circle; from a to δ is 10°, these two forming the top of the basin. It will be found very convenient to transport these dimensions in the mind's eye, as measuring rods of angular values in other portions of the heavens.

39. On the other side of the Pole Star, about equally distant from it, and lying also on the Equinoctial colure, is another notable group, commonly called THE CHAIR. It is represented in the upper right hand corner of the map, and is the principal portion of the constellation CASSIOPEIA. Its circle of apparent revolution about the pole is of the same magnitude as that of the dipper, and it is alternately near the zenith, or near the horizon, twelve hours after that group has occupied the same position. The center of the Chair passes the meridian a few minutes after the Vernal Equinox (See. 34).

40. The Chair and the Dipper are always available to the observer in the northern hemisphere, and both are always above the horizon of the northern States. They furnish us with the elements of a very important surveying line — that from which the Right Ascensions of all the stars are measured. The perpendicular line on the map represents the northern portion of the Equinoctial colure (See. 36) which cuts the Equinoctial at right angles, at the points of its intersection with the Ecliptic (See. 18). The star marked β — Chaph — in the back of the Chair, and Megrez, are very nearly on this line, the Chair end of which runs through the Vernal, and the Dipper end through the Autumnal, Equinox. These two stars are nearly equidistant from the Pole, and, in addition, enable us to find its position with reference to the Pole Star, which the map shows to lie exactly on a line from δ in the Chair to ζ in the dipper. The following are their places for January 1st, 1875:

STAR.	NAME.	RT. ASCENSION.	DECLINATION.
β Cassiopeiæ	Chaph	0h. 2m.31s.	58°27'37"
δ Ursæ Maj.	Megrez	12h. 9m.14s.	57°43'36"
α Ursæ Min.	Alruccabah	1h.12m.55s.	88°38'35"

The Pole Star is therefore 1° 21' 25" from the Pole, towards the Chair, and makes an angle of 1h. 12m. 55s, or 18° 13' 45" (See. 26) with the colure, inclining towards δ Cassiopeiæ.

41. CASSIOPEIA — The Lady in her chair. All of this constellation, except the head, is shown on Map No. 1. Its greater portion lies east of the Equinoctial colure. It is situated about half way between the North Pole and the Equinoctial, and contains 55 stars visible under the most favorable circumstances, but those shown in the figure are all that the student will easily recognize. Schedir — α — a pale, rose-tinted star of the 3rd magnitude, in the bosom; Chaph — β — a white star of the 2nd magnitude, in the back of the Chair; and γ, a white star of the 3rd magnitude, in the lap, form a right angled triangle, whose sides are about 5° each. Opposite to α is κ, of the 4th magnitude, and a star of the 5th magnitude, just north of κ, completes the square; δ, of the 3rd magnitude, in the knee, and ϵ, of the 3rd magnitude, in the forward foot, 5° apart, complete the front outline of the figure; ι forms the southern corner of a similar square of stars, but smaller than those in the body. The head of Cassiopeia, containing ζ, of the 4th magnitude, and the body, are in the Milky Way, the course of which is shown on the map.

42. URSA MINOR — The Lesser Bear. The Pole Star, Alruccabah, is situated at the extremity of the handle of another dipper-shaped group, rather smaller, and composed of smaller stars, than that in Ursa Major (See. 36). The two stars, β, of the 3rd magnitude, and ζ of the 4th, forming the top of the dipper, lie nearly in a straight line between Alkaid and Chaph. The handle forms the tail of Ursa Minor, and the basin gives the position of the body; Kochab — β — the nearest to the great dipper, is in the shoulder, and with γ, of the 3rd magnitude, in the breast, 3° south east of β, forms the front end of the little dipper; they are pointers to ζ, of the 3rd magnitude in Draco; δ, of the 3rd magnitude, is about 4° south of α, and between

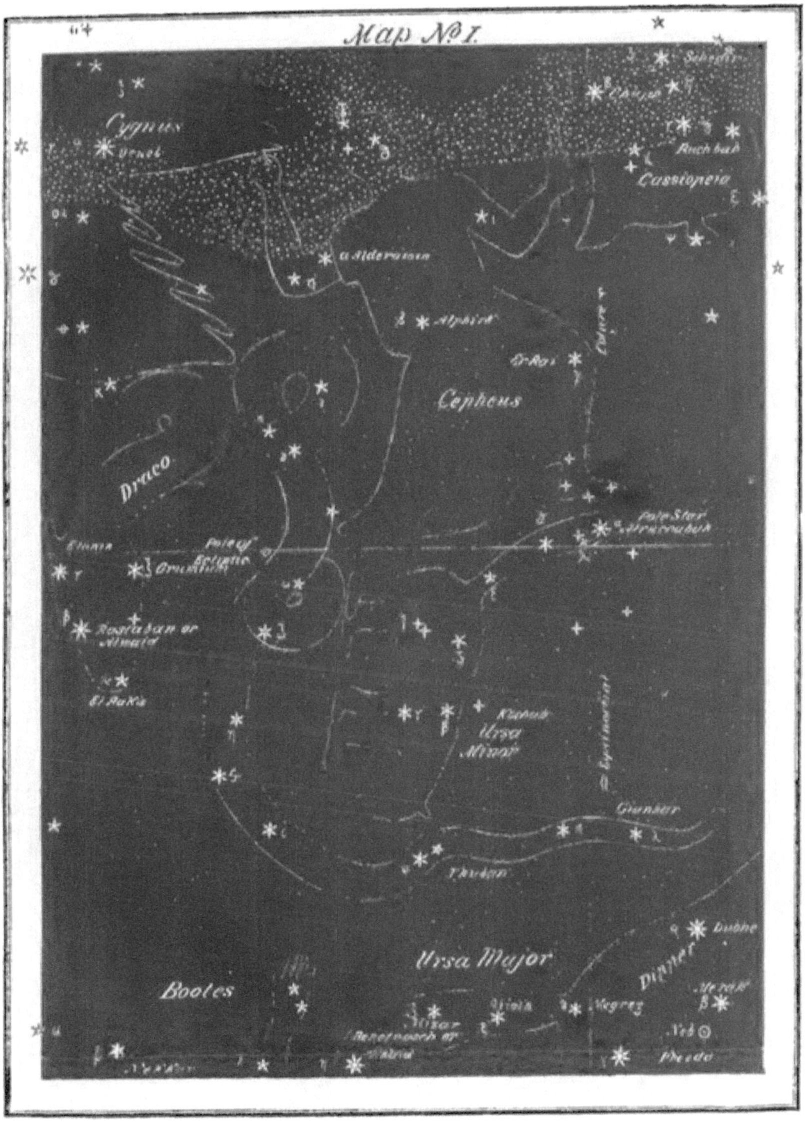

Map No. I.
Cygnus
Cassiopeia
Ruchbah
Caph
a Alderamin
Alphirk
Er Rai
Cepheus
Draco
Pole of Ecliptic
Pole Star
Ursa Minor
Kochab
El Rakis
Thuban
Ursa Major
Dubhe
Dipper
Merak
Megrez
Phecda
Bootes
Alioth
Mizar
Benetnasch or Alkaid

them are two contiguous stars of the 5th magnitude. The constellation contains 24 discernible stars.

43. CEPHEUS — Mythologically the king of Ethiopia, and husband of Cassiopeia, is represented on the globe, in regal state, with crown and sceptre. The constellation lies northeast of Cassiopeia, the feet near the pole, and contains 35 discernible stars, of which only six are prominent. A line from Schedir — α Cassiopeia — through Chaph, produced 20° further to the northeast, will nearly pass through two stars, 4° asunder, the nearest of which is Alderamin — α — a white star of the 3rd magnitude, in the left shoulder of Cepheus; 8° north of α is Alphirk — β — a white star of the 3rd magnitude, and these two form, with δ in the head, of the 4th magnitude, and another of the 4th in the right shoulder, a diamond square of stars, of about 8° on each side. Er Rai — γ — a yellow star of the 3rd magnitude, in the knee, lies nearer the pole, and nearly on the colure, at the vertex of a triangle with β and No. 1, and 11° distant from each. The position of the foot of Cepheus, very near the pole, is marked by four stars of the 5th magnitude, with one in the ancle nearer to γ. The head lies in the milky way.

44. DRACO — The Dragon. This constellation contains 80 discernible stars. When Cassiopeia is on the meridian above the pole, and the Dipper near the horizon, as in the map, a wedge-shaped group of bright stars, the point downwards, is in the northwest, about 35° above the horizon. It is nearly the same distance from the pole as are those two groups, and on the 18 hour circle; this circle, a portion of which is represented by a straight line on the map, is perpendicular to the Equinoctial colure, cutting the Ecliptic in the beginnings of ♋ and ♑ (See. 23), and the Equinoctial at 6 and 18 hours, forming the Solstitial colure. This group is the head of Draco, and circles around the pole 6 hours after the Dipper, and 6 hours before Cassiopeia, its Right Ascension being 18 hours. It may also be found by tracing a line from δ Cassiopeia through β Cephei, and another from Phecda through Megrez; it is nearly at the junction of these two lines, about 35° from the Pole. The two bright yellow stars of the 2nd magnitude, Etanin — γ, and Alwaid — β, with ξ of the 3rd magnitude, form nearly an equilateral triangle of 5° on each side, a line from γ through ξ, almost touching the pole; El Rakis — μ — in the nose, is 5° west of β, and forms another triangle with it and a small star lying between μ and ξ. The Solstitial Colure on which γ, and ξ — Grumium — lie nearly, passes through the poles of the Ecliptic, as well as those of the Equinoctial. The north pole of the Ecliptic is shown on the map, about 9° north of Grumium, and 23½° from the Pole Star, being 90° distant from every part of the Ecliptic Circle, as the pole of the earth's rotation is every where distant 90° from the Equinoctial. The word "Pole" always means "the pole of the earth's rotation," unless otherwise described as the Pole of the Ecliptic, or of some other circle; for every circle of the sphere has its own pole on the surface, as every circle on a plane has its own center.

45. The figure of the Dragon is a very irregular one, but it is easily traced in the heavens. From the head it proceeds eastward, nearly to ι Cygni, of the 4th magnitude, 12° from Etanin; then turning north, it winds round δ of the 3rd magnitude, 13° from ι, and thence curves irregularly round the Pole to the West, inclosing Ursa Minor, and taking in a circling line of several bright stars, including α, ι, and λ, all of the 3rd magnitude, near the Dipper, and between it and the pole; Thuban — α — a pale star near the head of Ursa Minor, and about half way between the pointers of that constellation and Mizar, the middle star in the handle of the Large Dipper, is remarkable as having been the Pole Star many ages ago, then occupying the place now held by Alruccabah, and being at one time nearer to the Pole than the present Pole Star now is. The Precession of the Equinoxes (Sec. 25) causes the stars near the pole to change their angular positions in Right Ascension more rapidly than those in the Zodiac — Alruccabah now increasing its right ascension by nearly 20¼ seconds of time yearly. The longitude is increased about 50¼" annually, the same in every part of the heavens. The places of the fixed stars, as given in this book, are those occupied January 1st, 1875, and are near enough to the actual places for observing purposes, for several years before and after that date.

46. The forward half of Draco now winds round the pole of the Ecliptic, and the latter half formerly included the pole star of the world. The shape was probably assigned to the stars it now embraces, by the first observers, long before the changes made in other portions of the heavens by the Greek mythologists, earlier, even, than the time of the Pharaohs — assigned there for the purpose of symbolizing the unity of the two systems of diurnal and annual motion, and furnishing an imperishable record of Astronomical research and patriarchal wisdom in the days when books and paper were unknown. The North Pole Star was then between the two Dippers, which were always over the polar regions. The ancient Chaldeans had even less knowledge of the character of the earth's surface near the pole than we have, and called it the "Country of the Bears;" "Under the Bears" became a well understood equivalent for "The Northern Lands." The idiom was adopted by other peoples, and has not yet been totally swallowed up in the stream of Time, though the Precession of the Equinoxes has carried the Great Dipper outside of the original track. It is, however, remarkable, that Thuban resigned his place to the principal star in one of the Bear groups, and Draco yet partially encloses both poles; still silently telling the old truth, but in new and less direct form.

47. A line from the foot of Cassiopeia through δ Cephei, and extended about 20° beyond δ, would nearly touch Deneb — α — a brilliant white star of the 1st magnitude in the constellation Cygnus, which crosses the meridian very near the Zenith of the Northern States. About 5° South, and a little West of Deneb, is γ of the 3rd magnitude, the two forming the shortest side of a triangle, with δ of the 3rd magnitude, nearer the head of Draco. The four stars lying South-east from Alkaid, in the lower left hand corner of the map, are in the Constellation Bootes: Nekkar — β — of the third magnitude, is 16° South-east from Alkaid, in the line of the Dipper handle.

DEFINE AND EXPLAIN (the figures refer to the sections):

36. The Dipper; its place; guide to Pole Star. 37. Changes in position. 38. Pointers; Alruccabah; measures of 10' and 25°. 39. The Chair; opposite what? 40. Equinoctial Colure; Chaph and Megrez; Pole Star and Pole. 41. Cassiopeia; The Square; Schedir; knee and foot; Milky Way. 42. Ursa Minor; small Dipper; Alkaid to Chaph; shoulder; pointers to Draco. 43 Cepheus; position; Alderamin; diamond figure; Er Rai; the foot. 44. Draco; the head; Solstitial Colure; Six hours after the Dipper; Equilateral triangle; Etanin; Grumium; Pole of Ecliptic; Poles of other circles. 45. Course of Draco; Thuban; former Pole Star; precession, effect near the Poles; places of stars in this book. 46. Origin of Draco; the two Dippers; early records. 47. Deneb; Nekkar.

MAP NO. II.

48. This map is connected with the first by the Constellation Cassiopeia, represented near the middle of the upper margin of the map, in an inverted position; the feet towards the pole, with Chaph on the curved line, in the right half of the Map, representing the Equinoctial Colure, and crossing the Equinoctial Circle at its intersection with the Ecliptic, in the first part of ♈ (Aries) near the lower margin of the map.

49. The line from Megrez, through the pole to Chaph, if continued 30° farther South, will pass nearly through a bright white star of the 1st magnitude — α — named Alpheratz, in the head of Andromeda; the line continued 14° still further, will pass a little West of Algenib — γ — a yellow colored star of the 2nd magnitude, in the wing of Pegasus; 14½° still further South the line crosses the Vernal Equinox, which is easily located by remembering that it is a very little farther South of Algenib than that star is South of Alpheratz.

50. These two stars form the Eastern side of a large square, with Markab, a white star of the 2nd magnitude, and Scheat, a deep yellow star, also marked α and β, in the constellation Pegasus. The last named two are just West of the circle of 23 hours of Right Ascension, and 12¾° apart. They are shown near the right margin of the map. Alpheratz and Algenib are on the meridian, about midnight on the 23d of September (page 12); 10 P.M., on October 23rd; 8 P.M., on November 22nd; and 6 P.M. on December 21st. They are always near the meridian at the same time that the Dipper is near the horizon, as in Map No. 1; Alpheratz being then 10° or 15° South of the Zenith, in the Northern States. Markab and Scheat Pegasi come to the meridian one hour earlier. The quadrangle is known as THE SQUARE OF PEGASUS, and its position in the heavens should be made familiar, as it is of great service in locating the Ecliptic and Equinoctial, which, in that part of the heavens, have few other notable stars near them.

51. A line drawn diagonally from Markab to Alpheratz, and prolonged to the Northeast, will pass a row of bright stars noted on the Map as Mirach, Almaach, and Mirfak. The middle three of the five form nearly the medial line of Andromeda, — α in the head, β, a yellow star of the 2nd magnitude in the girdle, and γ, an orange colored star of the 3rd magnitude, in the right foot. The last of the series is α, in the breast of Perseus; the other four are remarkable as lying nearly on successive hour circles, their Right Ascensions being as follows: α Pegasi, 22h. 58m. 32s; α Andromedae, 0h. 1m. 16s.; β Andromedae, 1h. 2m. 44s.; γ Andromedae, 1h. 56m. 14s. The gradual narrowing of the spaces between the hour circles, as they approach the Pole, is shown on the map; it lessens the angular distance as measured on the great circle of the sphere, in proceeding Northward. The following are the distances between these mile-stones in the firmament: Markab to Alpheratz 21°, Alpheratz to Mirach 14°, Mirach to Almaach 13°, Almaach to Mirfak 16°.

52. A line from ε, in the foot of Cassiopeia, South through Almaach, will pass Hamal 19° beyond, a yellow star of the 2nd magnitude — α — in the head of Aries. On the same line, 26½° South from Hamal, lies Mira, in Cetus — a variable

18   ASTRONOMY.

star, marked on the map as of the 2nd magnitude. The circle of 2 hours of Right Ascension, leaves ε Cassiopeiæ a little to the West, and passes through Hamal, passing, also, β Trianguli, a star of the 4th magnitude, about half way between Almaach and Hamal, and runs a little East of El Rischa, a pale, greenish colored star of the 3rd magnitude, which is 7° Northwest from Mira, and nearly 21° South of Hamal. The three last named are important; the Equinoctial passes nearly midway between El Rischa and Mira, and the Ecliptic nearly midway between El Rischa and Hamal.

53. Mirfak — α Persei — (See 51), a brilliant lilac star of the 2nd magnitude, is 18° Southeast of ε Cassiopeiæ; nearly midway between them is ζ Persei, in the sword arm, and between α and χ is γ of the 3rd magnitude, near the shoulder. This line is further prolonged to δ of the 3rd magnitude, 11° from ζ. South of γ 12°, and 8½° Southwest from δ, is Algol — β Persei — a dull white star of the 2nd magnitude, and marking the location of the third hour circle. A line from Algol to δ makes a right angle with a line from δ to γ. Northeast of δ and ζ, and forming with them a trapezium very narrow at the top, are two stars of the 4th magnitude, the principal ones in Camelopardus — a constellation not important enough to be represented on these maps. The Milky Way runs Northwest through the upper part of Perseus, then turning West takes in the body of Cassiopeia, and, passing nearly due West to the circle of 22 hours of Right Ascension, turns South towards Cygnus (Map xiii).

54. ANDROMEDA. — Mythologically the daughter of Cepheus and Cassiopeia, and bound to a rock — is represented on the globe with arms outstretched and fetters on the wrists. It contains 66 discernible stars; Alpheratz — α — is in the head, Mizach — β — in the girdle, and Almaach — γ — in the right foot. (See. 51.) The right hand nearly touches Algenib. Nearly midway between α and β is δ of the 3rd magnitude, in the right breast, forming one of a line of four stars, of which the most Southerly — ζ of the 4th magnitude — is in the right elbow, and the Northern one — π of the 4th magnitude — is in the center of the breast. β is the most Southerly of three in the girdle. Nearly half way between γ and Cassiopeia are No. 51 of the 3rd magnitude, and No. 51 of the 4th magnitude, which mark the position of the left foot. The stars of the 4th magnitude — ν, κ, and ο — Northwest of the figure, belong to Andromeda; ο is in a line with, and North from, the West side of the Square of Pegasus. The Nebula 2° West of ν, in the girdle, is visible to the naked eye. 10° Northwest from ο is α Lacertæ, of the 4th magnitude, the only star of any note in the constellation called THE LIZARD.

55. PISCES — The Fishes. This constellation lies South of Andromeda and the Square, crossing the Ecliptic, and lying between the circles of 23 hours and 2 hours of Right Ascension; it is represented as two fishes, connected by a long riband, and contains 113 discernible stars, but very few large enough to be easily found with the naked eye. PISCES OCCIDENS — The Western Fish — lies North of the Equinoctial, and almost parallel with it, and West of the Equinox; it contains six stars, as marked — β in the head, of the 5th magnitude, being 11½° South of Markab, in a direct line with the Western side of the great Square. Near to β is γ, the most Westerly of a row of three almost equidistant stars of the 4th magnitude, nearly parallel with the Southern side of the Square, and 8° North of the Ecliptic. PISCES BOREALIS, the Northern Fish, is parallel with the circle of 1 hour in Right Ascension, the head reaching to the girdle of Andromeda, and the tail to the Ecliptic. It may be located by tracing the line of small stars from ζ, in the elbow of An-

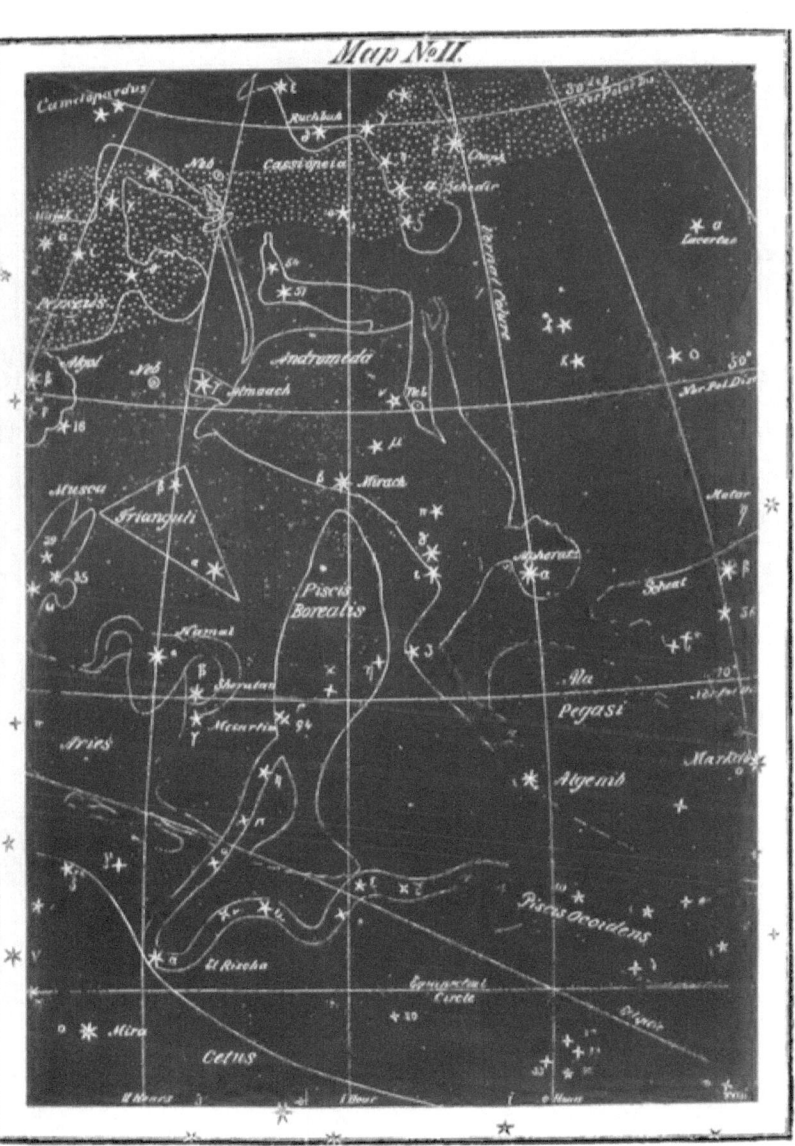

dromeda, Southeast to El Rischa (Sec. 52) — the third, marked ρ and 94, being visible as two stars of the 5th magnitude. The three following these are in the riband, the Ecliptic passing midway between ο and π. At El Rischa — α Piscium — the riband turns West in a waving line to the tail of the Western Fish, crossing the Ecliptic on the first hour circle, the intersection being between ε and ς. The small square of stars South of the Equinox belongs to Pisces; those in the extreme lower right hand corner are in Aquarius. The long line of prominent stars, Southeast of Pisces, are in Cetus.

56. ARIES — The Ram. This Constellation — mythologically the Ram that bore the golden fleece — is East of Piscis Borealis, being represented as recumbent on the Ecliptic, with one fore leg stretched out across the node of the riband of Pisces. It contains 66 discernible stars. From Hamal — α — of the 2nd magnitude (See 52) in the middle of the forehead, nearly 4° West, is Sheratan — β — a pearly white star of the 3rd magnitude; and 1½° South of β are two stars of the 4th magnitude — γ¹ and γ² — only 9' apart, and shown as one on the map. They are named Mesarthim. These are all the prominent stars in Aries. 16° a little South of East from α is δ of the 4th magnitude, in the hinder part of the animal, and π, ς, ζ and τ — all of the 5th magnitude — near δ, complete the ill defined figure. The last named stars are shown in Map No. iv.

57. MUSCA — The Fly. Northeast from Hamal, and about the same distance from Algol, Southwest, is a group of three stars — one of the 3rd magnitude, and two of the 4th — usually catalogued as in Aries, but known as the Fly. There is another Musca near the South Pole, a very unimportant constellation.

58. TRIANGULI — The Triangles. The Star β Trianguli (See. 52), of the 4th magnitude, lies at the Northern angle of a figure formerly marked with two triangles — now but one ; α — a yellow star of the 3rd magnitude, 7° from β, and the same distance from Hamal towards Cassiopeia, — marks the Western Angle, and forms the vertex of an obtuse isosceles triangle, whose base, bounded by β Trianguli and α Arietis, is on the circle of 2 hours of Right Ascension.

DEFINE AND EXPLAIN (the figures refer to the sections):

48. Connection of Maps I. and II. 49. Equinoctial Colure extended ; Alpheratz ; Algenib. 50. Square of Pegasus. 51. Alpheratz to Mirfak ; Shining milestones. 52. Hamal ; El Rischa ; Mira. 53. Mirfak ; Algol ; Camelopardus ; Milky Way. 54. Andromeda ; Alpheratz ; Miracli ; Almaach ; remarkable Nebula ; Lacerta. 55. Pisces, Western and Northern ; elbow of Andromeda to El Rischa ; Aquarius ; Cetus. 56. Aries ; Hamal ; Sheratan ; Mesarthim ; Southeast from Hamal. 57. Musca. 58. Trianguli.

MAP NO. III.

59. This map covers nearly the same portion of the heavens as the preceding, but is differently projected, being arranged with the Ecliptic as the principal line, the curved lines running up and down the map being circles of Latitude, meeting in the Ecliptic Poles (See. 44). The Equinoctial runs obliquely; the Square of Pegasus is in the upper right hand corner. Maps III. to X., inclusive, are constructed on this plan, enabling the student to trace out the paths of the members of the Solar System, and to refer them to the Solar Path, more readily than could be done on an Equinoctial projection. (See. 27.)

60. THE ZODIAC. The constellations which lie on the Ecliptic line are called the Zodiac, the word meaning "the circle of animals;" but it is now used also to indicate a band of 16° in breadth, 8° on each side the Ecliptic. Some 2,800 years ago, the Zodiacal list was remodelled, and the twelve signs received the names they now bear, (See. 23.) Each of those signs was then occupied by a group of stars bearing the same name, as: Aries then occupied the first 30° East from the Vernal Equinox, and to Pisces was allotted the 30° next West of that point. The Precession of the Equinoxes (See. 25) has, since then, carried them back more than a whole sign over the constellations Pisces and Virgo, so that the stars then in the sign ♓ are now in ♈, and those then in ♍ are now in ♎. The number, twelve, was probably chosen because that was the nearest even representative of the number of new moons in the year.

61. The first group was probably called "Aries" because, at the time the Sun was among those stars the young of the flock were born: not because the ancients fancied a resemblance to the shape of a sheep among the stars. It was a symbolic mode of recording a fact (See. 46). So, the place of the Sun, during that part of the year corresponding to the period now included between the middle of February and the 21st of March, was called Pisces, because that space marked the fishing season. There was also, perhaps, an allegorical reason for the latter in the singular paucity of notable stars in that part of the heavens, which might be considered as indicative of the poverty of fish diet.

62. CETUS — The whale. This constellation occupies the Southern half of the map, lying South of Pisces and Aries, nearly parallel with the Ecliptic, the head on the Equinoctial, reaching into the middle of the sign ♓; the tail extends to 15° or 18° West of the Equinox. Cetus occupies nearly three hours in passing the meridian, its center being about half an hour later than Cassiopeia. It is seen in the Southern part of the heavens, about 30° from the horizon, about the same times that the Dipper is visible near the Northern horizon. It contains 97 discernible stars, including several of prominence. Mira — α Ceti — in the neck (See. 52) 7° Southeast from El Rischa, and in a direct line with the stars in that part of the riband nearest to Piscis Borealis, is a variable star, being sometimes of the 2nd magnitude, and at others so small as to be scarcely visible. It is on the circle which divides the signs ♈ and ♉ and passes through β Andromedæ. Nearly 13° East of Mira is Menkar — α of the 2nd magni-

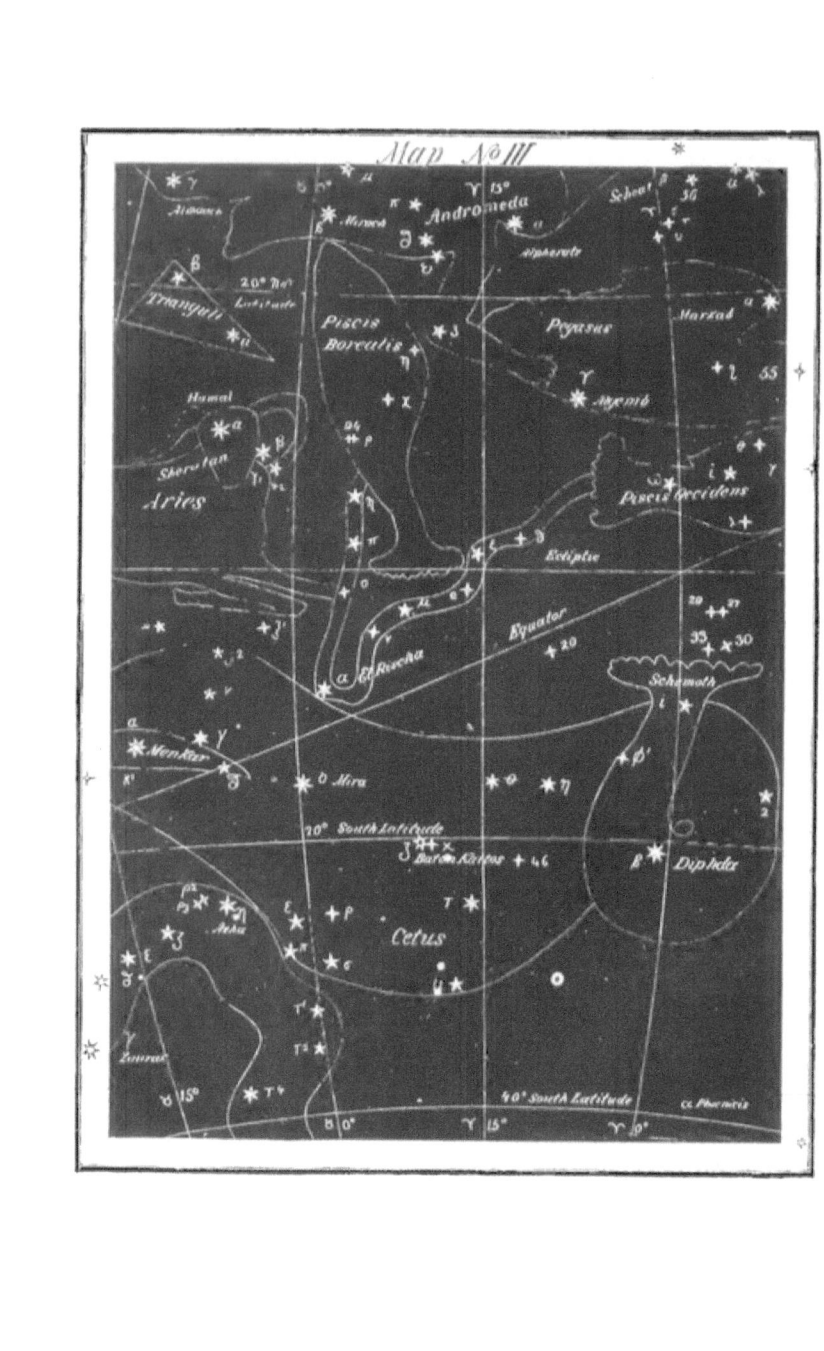

tude in the jaw; δ, of the 4th magnitude, is between o and a, and γ, of the 3rd magnitude, is a little further North, and in a direct line between Menkar and El Rischa; γ is the root of a line of three stars in the direction of β Arietis; 15° South of γ, and about 10° from Mira, is a trapezium of small stars in the breast. A line from Menkar, through Mira, to the tail of Cetus, will take in θ and τ, both of the 3d magnitude; and nearly half way between θ and the trapezium, is ζ of the 3d magnitude, named Baton Kaitos. Diphda — β — of the 2nd magnitude, is very near the circle dividing the Signs ♈ and ♓, with 21° of South latitude; it is 40° distant from Menkar, nearly on the same line with Mira: Alpheratz forms the vertex of a triangle, having a and β Ceti at the angles of the base.

63. PHŒNIX. A constellation called the Phœnix lies South of Cetus; it contains 13 discernible stars. a, of the 2nd magnitude, in the neck, is shown on the margin; it is situated 24° South of Diphda, and just East of the Equinoctial Colure, on a line with Megrez, Chaph, Alpheratz, and Algenib; β, of the third magnitude, is 8° Southeast of a, and on the circle of 1 hour of Right Ascension; γ, of the 3rd magnitude, is 12° East of a, and 5° Northeast of β.

64. ERIDANUS — The River Po. The stars under the neck of Cetus, in the lower left hand corner of the map, are in the constellation Eridanus — a long, straggling assemblage of stars. The upper portion of the stream, shown on this map, is at the bend which connects the Northern and Southern portions of the constellation. The Northern part is shown entire in the next map; the Southern stream stretches far towards the South Pole. (See Map No. xiv.) A line from the head of Aries, through Menkar, and produced 22° farther, will pass through Zaurak — γ — of the 2nd magnitude, taking in δ, of the 3rd magnitude, on the way; from δ a line of prominent stars stretches half way to Mira. Below the trapezium of Cetus are τ^1 and τ^2, of the 4th magnitude, and lower, nearer Zaurak, is τ^4, of the 3rd magnitude. Beyond these is the largest star in the constellation — a — of the 1st magnitude, named Achernar; it is too far South to be shown on the map, or seen in the Northern States, being 58° South of the Equinoctial; it is 19° Southeast from a Phœnicis, and 51° a little West of South from Zaurak. A little east, and 21° South of Zaurak, and 18° Southeast of τ^4, is o^1, with o^2 1° East of the latter. Theemin — o^3 — is 5° Northeast from o^1, in line with Achernar. o^5 is of the 4th magnitude, the other two are of the 3rd magnitude.

DEFINE AND EXPLAIN (the figures refer to the sections):

59. Map No. III.; the Square. 60. The Zodiac; ancient and modern; changes; effect of Precession. 61. Aries and Pisces. 62. Cetus, extent; Mira; Menkar, line to Sheratan; Diphda. 63. Phœnix; with Megrez. 64. Eridanus; Zaurak; Achernar; Theemin.

MAP NO. IV.

65. This map shows a section of the heavens lying immediately East of that shown in Map No. III.; and includes γ, δ, ε, and ζ, Eridani; α Ceti, β Trianguli, and γ Andromedæ, in common with it; also Musca and the head and body of Perseus in common with Map No. II. It takes in a section from the middle of the Sign ♉, to a little past the first degree of ♋, the Colure of the Summer Solstice. The Equinoctial is represented running obliquely, from 15° to 23½° South of the Ecliptic. The Right Ascension of the center of the map is about 4½ hours, and it passes the meridian 1½ hours after the Vernal Equinox. This portion of the heavens is more brilliant than any other of equal extent.

66. PERSEUS — Mythologically, the deliverer and subsequent husband of Andromeda, and slayer of the Gorgon — is represented on the Globe as in the act of making a back stroke with his sword at the head of Medusa, which is behind him. The Sword is shown in Map No. II. The constellation contains 59 discernible stars. Of the line of stars, δ, γ, α, and δ (See 53), only the Southern three are shown on this Map; θ, of the 4th magnitude, in the neck, is 7° from γ towards Almaach, and the same distance from α Persei, forming the vertex of an isosceles triangle with α and γ; ι, of the 4th magnitude, lies between θ and α. About 6° beyond δ from Almaach, is μ, of the 4th magnitude, in the left knee, and above it is λ of the 4th in a line beyond α with ε and θ. A line of stars, curving Southward, round Algol, several degrees to the East of that star, form the hinder leg of Perseus; ζ of the 3rd magnitude, the most Southerly, being 19° from Mirfak. Two stars of the 4th magnitude are here shown near Algol, in the head of Medusa, which will aid in locating it (See 53). The Milky Way includes the whole of the body, and the forward leg, of Perseus.

67. TAURUS — The Bull. This constellation is now the third in order of the Zodiacal band, lying directly East of Aries, the principal stars being in the Sign ♊. Only the front portion of the animal is represented on the globe, but the great length of the horns compensates for the curtailment of the body. It contains 141 discernible stars. South of ζ and ο in the foot of Perseus, 8°, and 19° below Algol, is a well known group called the Pleiades, containing Alcyone — η — a greenish yellow star of the 3rd magnitude, and five others, four of which are of the 5th magnitude; the stars in this group are a little nearer together than indicated on the map. West of the Pleiades, 8°, is the group of small stars in the hinder part of Aries (See 56). Southeast from the Pleiades 14°, and 32° below Algol, through σ Persei, also 26° from Menkar, is a pale rose colored star of the 1st magnitude, called Aldebaran — α — in the eye of the Bull; and 3° above it, in the direction of ζ Persei is ε, of the 3rd magnitude. α and ε form the top of a prominent V-shaped cluster called the Hyades, with γ of the 3rd magnitude in the angle, pointing towards Menkar. A line of smaller stars running from just below the V towards Menkar, with a lower line, nearly parallel with these, locate the legs of the Bull. A line from Menkar through Aldebaran, produced 15° farther, will strike the Ecliptic be-

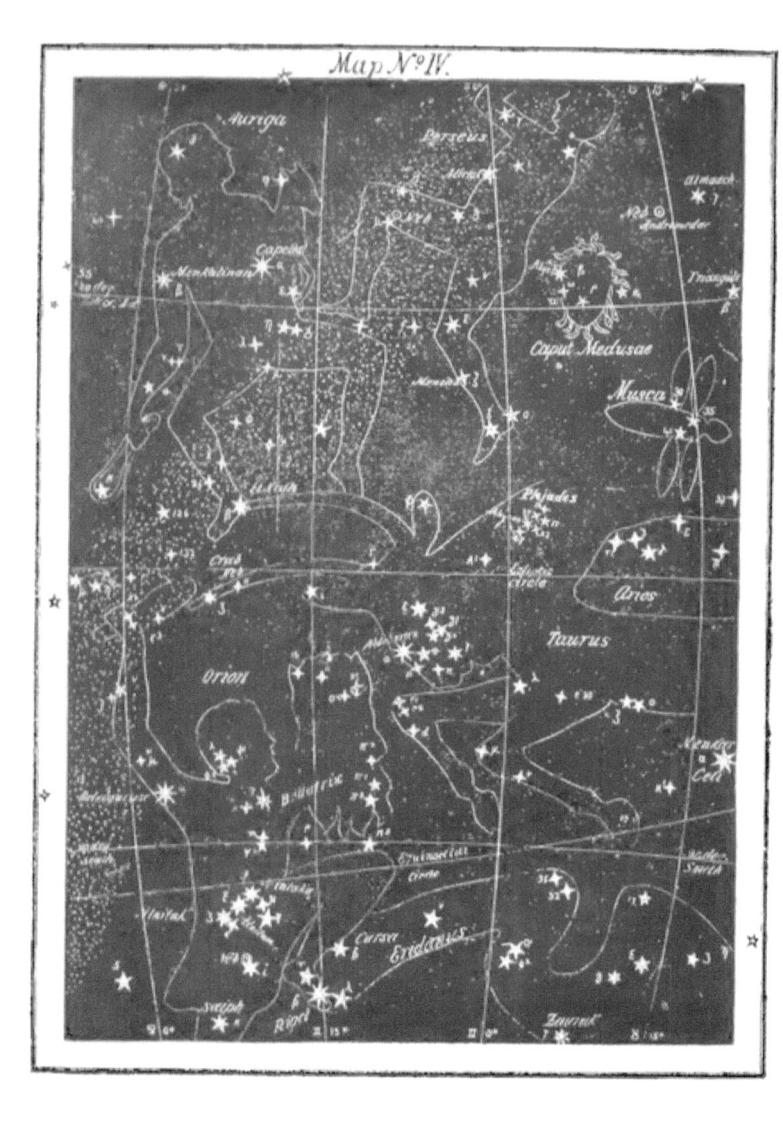

tween two prominent stars at the tips of the Bull's horns, situated 8° apart; El Nath — β — the most northerly, is a brilliant white star of the 2nd magnitude; the Southern one is ζ, of the 3rd magnitude. Three small stars, nearly in a line with those in the lower foot of Perseus, mark the tops of the head and ears. This constellation is supposed to have been named to mark the time of the year when the young of cattle are born.

68. AURIGA — The Wagoner. This constellation is represented on the globe as a mis-shapen man, with a goat under one arm, and a bridle in the other hand. The constellation lies East of Perseus, and North of the horns of Taurus, with the left foot resting on the tip of the North horn — El Nath — β Tauri — being common to both. The circle of 6 hours of Right Ascension forms its Eastern boundary; it contains 66 discernible stars. A line 18° north from El Nath will strike α, a brilliant white star of the 1st magnitude, in the right shoulder, and named Capella, the she goat. α Aurigæ is also 35° East of Almaach, and 8° farther East is Menkalinan — β — a yellow star of the 2nd magnitude in the left shoulder; θ, of the 4th magnitude, is 8° South of β, in the left arm; δ, of the 3rd magnitude, in the head, is 10° North of β, in line with θ. A line from δ through α, produced, will pass between the Hyades and Pleiades, taking in its course η and ζ in the groin, both of the 4th magnitude, and passing ι, of the 4th magnitude, in the right foot; η and ζ are best memorized as forming the Northern and shortest side of a small trapezium; λ, of the 5th magnitude, is on the line between Capella and El Nath, the latter being also at the lower corner of a compressed square of five stars. Nos. 132 and 136, Southeast of El Nath, belong to Taurus; those East of Menkalinan belong to Auriga, but are generally spoken of as lying in the Telescope of Herschel.

69. ORION — The Hunter. South of Auriga, and Southeast of Taurus, is the brilliant constellation, Orion; represented on the globe with club upraised, and shield presented to meet the assault of the Bull. It contains 78 discernible stars. The most prominent feature in Orion is the Belt, sometimes called "The Yard Stick — a row of three stars of the 2nd magnitude, δ, ε, and ζ, forming a line of about 3° in length; Mintaka — δ — the most northerly of the three, is but 0° 23′ 28″ South of the Equinoctial, and 46½° South of α Aurigæ. At right angles with the line of the belt, and nearly equi-distant from it, are two stars of the 1st magnitude; Betelguense — α — an orange colored star in the left shoulder, 10° Northeast, and Rigel — β — a pale yellow star in the right foot, 9° Southwest of the belt. Near β, towards the belt, is τ of the 4th magnitude; 7° from both α and δ, and 15° above β, is γ, a pale yellow star of the 2nd magnitude, in the right shoulder; and 17° from γ, through the belt, is Saiph — κ — of the 3d magnitude, in the left knee; κ is also 8° East of β, γ is 10° South of Capella, and α is 38° South of Menkalinan, these four forming a long parallelogram, the diagonals of which would nearly cross in El Nath. A line from the belt, through α, will touch two small stars in the left elbow, nearly on the 6 hour circle, and North of these are three small stars, marked χ, in the club hand. Three small stars in a triangular position mark the head, and are at the vertex of a triangle whose base is formed by the shoulder stars. The shield is formed round a curved line of eight small stars, with γ at the center of the curve, and the convex side towards Aldebaran. South 4° from ζ, in the belt, is a sextuple star marked ι, at the end of the sword, and close to it is the celebrated Nebula in the sword. The stars east of the club hand are in the constellation Gemini; those east of the body belong to Monoceros — The Unicorn — a constella-

tion composed of small stars, and not outlined on these maps.

70. Rigel is the middle one of three adjacent stars, forming an arc — τ Orionis, of the 4th magnitude, in the foot, and λ of the 4th beyond, with β — Cursa — a topaz yellow, of the 3rd magnitude, a little above the three, and equidistant from them; λ and β are in Eridanus (Sec. 64) at the origin of the Northern stream, which runs West from the foot of Orion, parallel with the Equinoctial, and just below it, till it comes under the feet of the Bull, and then describes a semi-circle with the crown of the arch towards the South, taking in γ, δ, and ε, and No. 17 — four important stars, almost in a right line, and shown on Map No. III. The Milky Way runs from Perseus, Southeast, through the lower part of Auriga, and crosses the Ecliptic at the beginning of the Sign ♊, just including the club hand of Orion within its Western edge.

71. HYDRUS — The Water Snake. This constellation lies near the South Pole, and contains 10 discernible stars. 5° South East of Achernar — α Eridani — is α Hydri, of the 3rd magnitude, and 16° farther, in the same direction, is γ, of the 3rd magnitude; 12° Southwest of γ is β, of the 3rd magnitude, 24° below Achernar, less than 12° from the South Pole, and only 1° degree East of the Equinoctial Colure on the same line with Alpheratz. South of α, 6°, is ζ, of the 4th magnitude, which forms an equilateral triangle with β and γ. (See Map No. XII.)

72. RETICULUM RHOMBOIDALIS. — The Rhomboidal Net. 20° East of Achernar, and 12° North of γ Hydri, is α, of the 3rd magnitude; β, of the 4th magnitude, is 4° Southwest of α, towards ν Hydri.

73. DORADUS — The Sword-Fish. This constellation contains 6 discernible stars, lying East and North of the Net. 8° North of α Reticuli, and 24° East of Achernar, is α, of the 3rd magnitude; β, of the 4th magnitude is 9° East of α Reticuli, and 12° below a star of the 1st magnitude called Canopus. γ, of the 4th magnitude, is 5° a little West of South from α Doradûs, in a direct line with β 14° farther.

DEFINE AND EXPLAIN (the figures refer to the sections)

65. Map No. IV, connections; Right Ascension. 66. Perseus; Sword; isosceles triangle; line in hinder leg; Milky Way. 67. Taurus; Pleiades; Alcyone; proportion; Aldebaran; Hyades; Bull's horns. 68. Auriga; El Nath; Capella; Menkalinan; Telescope. 69. Orion; Belt; Mintaka; Betelgeuse; Rigel; Saiph; the Shield; Sword nebula; Monoceros. 70. Cursa; Northern stream of Eridanus; Milky Way. 71. Hydrus; South Pole. 72. Reticulum Rhomboidalis. 73. Doradus.

MAP NO. V.

74. This map shows the relative positions of the principal fixed stars lying within a distance of about 30° East of Orion, and the Northern part of Auriga; including Canis Major and Lepus — constellations farther South than the range of the preceding map. It shows, on its Western side, the stars Betelgueuse, Menkalinan, θ Auriga, μ and ν Geminorum, and No. 5 Monocerotis, in common with the preceding map. The Right Ascension of its centre is about 7¼ hours, and it crosses the Meridian 7¼ hours after the passage of the Vernal Equinox. It is much less thickly dotted with prominent stars than the preceding map.

75. GEMINI — The Twins. This constellation is represented on the globe as two youths, seated closely together; they are named Castor and Pollux, from two stars of the same names, one in the head of each figure — marked α and β respectively. Gemini is the fourth in order of the Zodiacal constellations, is just East of the sixth hour circle, and contains 85 stars discernible with the naked eye. A line drawn Southeast from Menkalinan will pass through Castor 22° distant, and Pollux 5° farther. Castor — α — is really two white stars of the 3rd magnitude, but so near together as to be ordinarily seen as one. Pollux — β — is orange colored, of the 2nd magnitude, 34° Northeast of Betelgueuse, and located on the right angle of a triangle whose hypothenuse is included by Menkalinan and Betelgueuse. From the latter, 11° towards Pollux, is Alhena — γ — a brilliant white star of the 2nd magnitude, and 7° northwest of γ is μ, of the 3rd magnitude, only 0° 30' South of the Ecliptic, and 3¼° beyond the beginning of the Sign ♋. From μ, 6°, towards Castor, is Mebsuta — ε — a white star of the 3rd magnitude, lying at the Northern corner of a square formed by ε, μ, γ, and ζ; that square is prolonged by α and β into a parallelogram 20° long, and 5° to 6° in breath. Wasat — δ — of the 3rd magnitude, lies 13½° East of μ, and is but a few minutes South of the Ecliptic; γ and μ are in a line of four stars indicating the position of the feet, two smaller stars at the Northern end curving towards the West, near the club hand of Orion. Equidistant about 4° from β are κ and ι, making with it an equilateral triangle, the three forming, with δ, a cross 9° in length and 5° in breadth. Tejat — η — of the 4th magnitude, is 2° West of μ; and No. 1, of the 5th magnitude, 2½° West of η, is named Propus, the name having been also given to ι Auriga, of the 4th magnitude, in the bridle of Auriga, and nearly in line with Wasat and Mebsuta. This constellation was probably named Gemini, because of the twin-like character of the season during which the Sun traverses the space formerly occupied by this group; it has also been suggested that, as the figure was originally that of two kids, the name was given to mark the time when the young goats first appear.

76. TELESCOPIUM HERSCHELIUM — North of Gemini and East of Auriga, is a modern constellation, called the Telescope of Herschel (Sec. 68), containing several small stars, generally catalogued as belonging to Auriga; they form a small trapezium North of Castor, and a small triangle a little Southeast of Menkalinan. There are no important stars between these and the Pole.

77. Cancer — The Crab. East of Gemini are five stars of the 4th magnitude, which, with a few of the 5th magnitude, constitute all that is ordinarily visible of Cancer, the fifth in order of the Zodiacal twelve, though it contains 83 discernible stars. It is represented on the globe as a large Crab with numerous claws. A line from Castor, through Pollux, produced about 8°, will touch ς², just North of the Ecliptic, with η², of the 5th magnitude, near and above it ; 9½° further Eastward is δ, a straw colored star of the 4th magnitude, also just North of the Ecliptic. A little more than 3° North of δ is γ ; these are called The Aselli, or The Two Asses ; nearly midway between the Aselli, but a little West of the line joining them, is a nebulous cluster, visible to the naked eye on a favorable night, and called Praesepe, or the Beehive. Acubens — α — of the 4th magnitude, is 6° Southeast of δ. The four stars marked ρ, ε, λ, and ε, Northeast of Cancer, are in the head of Leo. Cancer, formerly occupying the sign of the same name, was probably so called because the Sun, when in that part of the heavens, had ceased his advance towards the North Pole, and appeared to be progressing with a crab-like motion — sidewise — parallel with the Equinoctial, and soon receded towards the South — the Crab being, till a recent date, said to walk backwards.

78. Canis Minor — The Lesser Dog. This constellation, containing 14 discernible stars, lies South of Gemini, and Southwest of Cancer. Procyon — α — a yellowish white star of the 1st magnitude, is 26° East of Betelgueuse, 18½° Southeast of Alhena, and 18° almost due South of Pollux. Gomeisa — β — a white star of the 3rd magnitude, is 4° Northwest from Procyon, the two being on a line parallel with one passing through the heads of the twins, and are nearly the same distance apart as are Castor and Pollux. These two are the only stars of any note in the constellation. South of Cancer and East of Canis Minor, is a bent line of four stars of the 4th magnitude — δ, ε, ζ, and θ — and South of these is Alphard — α — of the 2nd magnitude. These are in the head and first fold of Hydra. Alphard is 8° South of the Equinoctial, 31° Southeast by East from Procyon, and 44° Southeast from Pollux, in a line with Castor.

79. Canis Major — The Greater Dog. This is a prominent constellation, but lying so far South that it is never seen far above the horizon in the Northern States. It contains 31 discernible stars ; all the prominent ones are shown on the map, though the figure of the Dog is not given entire. Sirius — α — a brilliant white star of the 1st magnitude, in the mouth, is 27° from Betelgueuse, 25° below Procyon, and 34° below Alhena, with 16½° of South Declination. Mirzam — β — of the 2nd magnitude, in the paw, is 5½° West of Sirius, and γ, of the 4th magnitude — Muliphen — in the head, is 3° East of Sirius, all three on a line with Alphard. A line from the belt of Orion, through δ, and produced 11° farther, will pass through Adhara — ε — of the 2nd magnitude ; 10° from Sirius, in a line with Betelgueuse, and 5½ East of ε is η — Aludra — of the 2nd magnitude ; 5° above η towards δ, is Wesen — δ — forming a triangle with η and ε. No. 22 is just above ε towards δ. α, β, ε, and η, form a figure nearly approaching to a long parallelogram. Sirius is familiarly known as "The Dog Star."

80. Lepus — The Hare. This constellation lies directly South of Orion, and contains 19 discernible stars. Below Rigel 11°, and a little West of 9° South from Saiph, is Arneb — α — a pale yellow star of the 3rd magnitude. A little West of 3° South from α, is Nibal — β — a deep yellow star of the 4th magnitude ; 4° Southeast from β, and 5° below α, is γ ; these, with δ, in the Eastern corner, form a trapezium, very similar in size and shape to the dipper of Ursa Minor ; 7° Northeast

α, and 5° South of Saiph, is χ, with ζ and θ about 2° distant — one on each side.

81. COLUMBA NOACHI — Noah's Dove. This constellation, not shown on the map, lies South of Lepus, and contains 10 discernible stars. South of Arneb 17°, South of the belt of Orion 33°, and 23° Southwest of Sirius, is Phact — α — of the 2nd magnitude; 3° Southeast of α, is β, of the 3rd magnitude, and 1° Northeast of ε is γ, of the 4th magnitude, Southwest of α, 2°, is ε; these four stars make a small rhomboidal figure.

82. Two stars of the 3rd magnitude, 4° apart, lying East of Canis Major — ρ and ξ — are in the peak of the prow of Argo — the ship. Tureis — ρ — is 21° Southeast from Sirius, and 25° Southwest from Alphard. The scattered stars between the two dogs belong to the untraceable constellation Monoceros — The Unicorn. The Milky Way runs diagonally across the groups shown on Map No. V.; from the arm of Orion and feet of Gemini, between the Dogs, taking in the Unicorn, and prow of Argo.

A space of more than 60° in length from the foot of Cepheus to Cancer, and 25° in breadth, Northeast of Cassiopeia, Perseus, Auriga, and Gemini, is peculiarly bare of stars prominent enough to be readily recognized; it contains scarcely a star of greater magnitude than the 5th. For this reason the space is not represented on the map, although astronomers have divided its stars into two constellations. The Lynx occupies a space about 35° by 17°, North of Gemini and Cancer, and East of Auriga. It contains but two stars of the 4th magnitude; these are situated 17° a little East of North from the Aselli (See. 77). The space North of Perseus, Auriga, and the Lynx, is filled up on the globe by the long figure of the Camelopard; its hinder hoofs nearly touch the head of Auriga (See Map No. IV.), and its head rests on the circle of 12 hours Right Ascension, a few degrees from the Pole. Camelopardalus contains four stars of the 4th magnitude, enumerated as Nos. 2, 3, 9, and 10. Nos. 2 and 3 are about 1° apart, and 6° Northeast from γ Persei (See. 53). East 12° from these, and 15° above Capella, is No. 10; shown on the margin of Map No. IV. North 6° from No. 10, in line with Capella, and 12° from Nos. 2 and 3, is No. 9. These four stars form an isosceles triangle, whose base is about half the height, its broken apex being directed towards the head of Perseus, and its Southern side parallel with the Equinoctial.

DEFINE AND EXPLAIN (the figures refer to the sections):

74. Map No. V.; Connections. 75. Gemini; Castor and Pollux; Alhena; Mebsuta; Wasat; Propus. 76. Telescopium Herschelium. 77. Cancer; Aselli; Præsepe; Acubens. 78. Canis Minor; Procyon; Gomisa; Alphard. 79. Canis Major; Sirius; Mirzam; Muliphen; Adhara; Aludra; Wesen. 80. Lepus; Arneb; Nihal. 81. Columba Noachi; Phact. 82. Argo; Tureis; Milky Way. The Lynx; Camelopardalus.

MAP NO. VI.

83. This map represents the constellations Leo and Ursa Major, both marked by many prominent stars; also Leo Minor, the Head of Hydra, and one of the Hounds of Bootes. It includes a section of the Ecliptic next East of that shown in the preceding map, but represents principally stars North of the Ecliptic. The middle of the lower part — the beginning of the Sign ♍ — has about 10 hours of Right Ascension, but where the center line cuts the Dipper the Right Ascension is about 12 hours, the hour circles being perpendicular to the Equinoctial, which is represented by the oblique line in the lower left. The map contains the head of Hydra, part of Cancer, and four stars in the head of Leo, in common with the one preceding it.

84. URSA MAJOR — The Great Bear. This constellation lies nearly midway between the Equinoctial and the Pole, and contains 87 discernible stars, including the Dipper (Sec. 38), which is situated in the rear half of the body and the tail. The Dipper stars are named, nearly in the order of their Right Ascensions, thus: α — Dubhe — a yellow star of the 1st magnitude; β — Merak — a greenish white, of the 2nd magnitude; γ — Pheeda — a topaz color, of the 2nd magnitude; δ — Megrez — a pale yellow, of the 3rd magnitude; ϵ — Alioth, of the 3rd magnitude; ζ — Mizar — a brilliant white of the 3rd magnitude; η — Alkaid, or Benetnasch — a brilliant white, of the 2nd magnitude. The small star above Alioth is named Alcor. The position of the nose of the Great Bear is marked by $\omicron$, of the 4th magnitude, outside the margin of the map; $\omicron$ is $18\frac{1}{2}°$ from Dubhe, in a direct line with Alioth and Alkaid. Talita — ι — of the 3rd magnitude, in the fore foot, is $20°$ from Merak, in a line with Megrez; θ, of the 3rd magnitude, is $6\frac{1}{2}°$ above ι in the same direction. The hind foot is marked by ξ and ν, both of the 4th magnitude, ξ being $31°$ Southeast from ϵ. Half way between ξ and β is ψ, of the 3rd magnitude, in the hind quarter; χ, of the 4th magnitude, is Northeast of ψ, and about equidistant from it and γ. Nearly on a line with ξ and ι, and midway between them, are μ and λ, both of the 3rd magnitude, and marking the place of the advanced hind foot. Nos. 23 and 29, in the neck, are easily located on lines from α to α and β.

85. CANES VENATICI — The Hounds of Bootes — are East of Ursa Major; the constellation contains 25 visible stars. A line from Dubhe, through Pheeda, produced $19°$ farther, would touch α, of the 2nd magnitude, and named Cor Caroli — the "Heart of Charles." The star is located in the heart of one of two hounds named Asterion and Chara, which, under the guidance of Bootes, were supposed to follow the two Bears around the Pole. The hounds are technically known as Canes Venatici, but only Chara is shown on our maps. α and Alioth are pointers to the pole.

86. LEO — The Lion. This is now the sixth in order of the Zodiacal Constellations; it contains 95 discernible stars, and is situated Eastward of Cancer, but is best located independently, or by reference to Ursa Major. It is represented on the globe as a huge animal with flowing mane, open mouth, and long tail; the body just North of the Ecliptic, and the hinder foot dipping below the Equinoctial. Regulus — α — a nearly white star

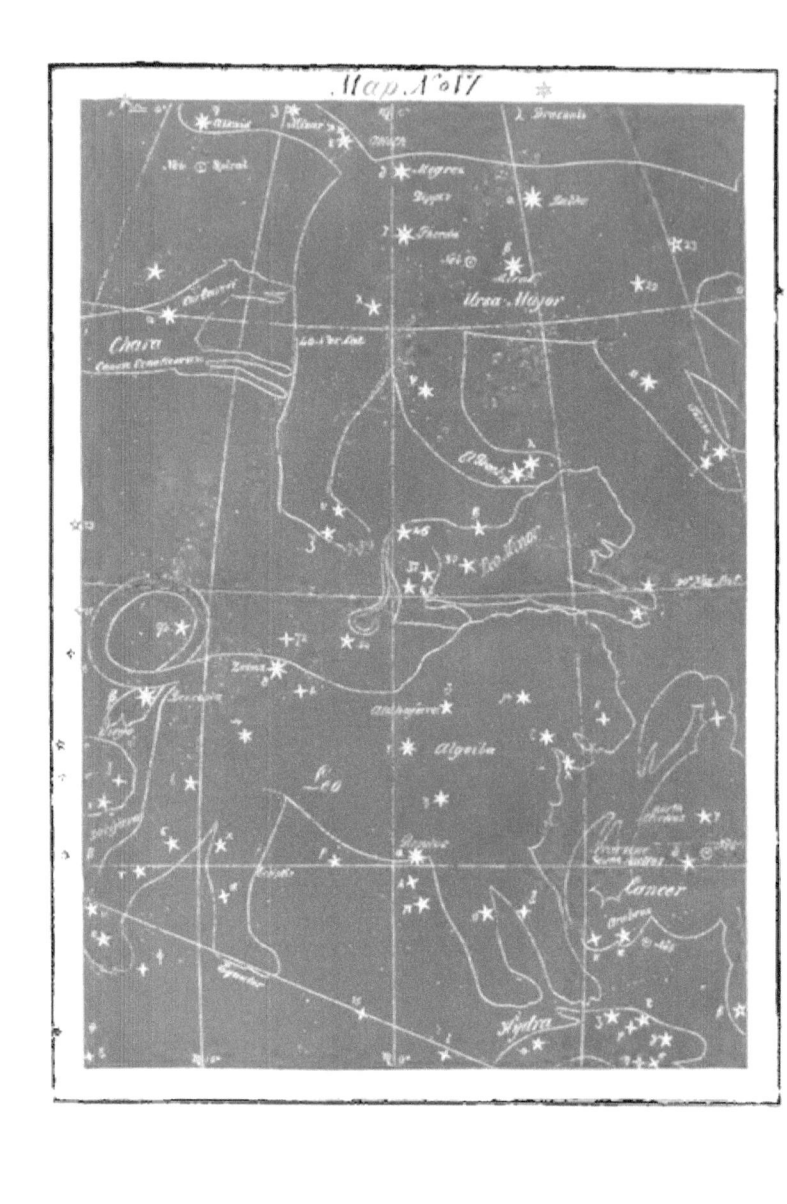

of the 1st magnitude, in the breast, called also "Cor Leonis," or the "Lion's Heart," is situated only 0° 28′ North of the Ecliptic, in 28° 5′ of the Sign ♌, and 22½° Northeast by North from Alphard; 3° North of α is η, of the 3rd magnitude and 8° 21′ distant from α is Algeiba—γ—a bright orange colored star of the 2nd magnitude, the line from which to α is perpendicular to the Ecliptic, and points directly to Pheeda and Megrez, in the Dipper; the distance from Regulus to Megrez is 41° 11′; η Leonis is equidistant from α and γ, towards the West, and forms, with α, the handle, of the well known Sickle of Leo, which curves round to the West, forming the arch of the neck, and ends in the lower jaw. The Sickle consists of γ, η, γ, ζ of the 4th magnitude, μ and ε, of the 3d magnitude, and λ of the 4th magnitude; κ of the 5th magnitude giving the limit of the nose about 4° North of λ. East from γ 13°, is Zozma—δ—a pale yellow star of the 2nd magnitude, in the back—hence sometimes called "Dorsa Leonis;" δ lies at the Northern end of a line of six stars, running due South, 23° in length, which forms the axial line of the hinder leg, and runs below the Ecliptic; the terminal star in the foot — e — of the 4th magnitude, is backed by σ, of the 4th magnitude, 2° distant, in direct line with a star of the 3rd magnitude, marked β, named Zavijava, and belonging to Virgo. The line from δ extended 7° below e, will pass through θ, of the 4th magnitude, in the Crater. A line from γ through η, will pass ω, of the 4th magnitude, and ξ, of the 5th magnitude, in the fore legs. Acubens, in Cancer, is about 8° West of the latter. The head of Hydra is Southwest of ω and ξ. A line from Regulus to σ, of the 4th magnitude, the fourth star in the hinder leg, will pass near ρ and χ, both of the 4th magnitude — the latter in the first hind leg. East of α, 25°, and 10° Southeast from δ, is Denebola — β — a bluish star of the 2nd magnitude, in the tail, and 13° above β Virginis, in direct line towards Pheeda. The stars East of Leo belong to the constellation Virgo. The Sun was formerly among the stars of Leo during the latter part of July and the first two thirds of August. They were named after the Lion, probably as a symbolism to indicate the raging heat of that portion of the year. The Sickle is now with the Sun in the harvest season.

87. LEO MINOR — The Lesser Lion. This constellation lies between Leo and Ursa Major. It contains 53 discernible stars, but only 5 of any note; these are all of the 4th magnitude, and form a trapezium which lies just below the hinder feet of the Bear, and a little farther above Leo, in a direct line between the Sickle and the Dipper. West of the trapezium and North of the Sickle, are two stars of the 4th magnitude, in the fore paws, but generally catalogued as belonging to the Lynx (See. 82).

DEFINE AND EXPLAIN (the figures refer to the sections):

83. Map No. 6; Contents; Connections. 84. Ursa Major; Dubhe to Alkaid; Alioth; Alcor; the nose; Talita; the feet. 85. Canes Venatici; Cor Caroli; the hounds; Pointers. 86. Leo; Regulus; Algeiba; the Sickle; Zozma to e; Acubens; Denebola. 87. Leo Minor; trapezium.

MAP NO. VII.

88. This map represents that portion of the Zodiac lying next east of Leo, but extends farther South than the map preceding. The Right Ascension of its center is about 13 hours; of the North portion more, of the South part less, than 13 hours. The stars shown near the center are on the meridian 1 hour after the Dipper occupies the meridian *above* the Pole (Sec. 37). The crossing of the Ecliptic and Equinoctial circles at the Autumnal Equinox, is shown to the right of the center. The map contains Cor Caroli, Denebola, with the stars in the hinder foot of Leo, the head of Virgo, and the West side of Crater, in common with Map No. VI.

89. VIRGO — The Virgin. This constellation is represented on the globe as a winged female, located parallel with, and principally North of, the Ecliptic, with the Equinoctial running diagonally through the figure; she has a bundle of grain in her hand, supposed to have been gleaned in the harvest field. It contains 110 discernible stars. Zavijava — β — a pale yellow star of the 3rd magnitude, is 13° South of Denebola, and near the Ecliptic at the tip of the lower shoulder; 18° nearly East from Denebola, in line with δ Leonis, is Vindemiatrix — ϵ — of the 3rd magnitude; 8° below ϵ, 16° East of Zavijava, and nearly at the same distance North of the Equinoctial, is δ, a golden yellow star of the 3rd magnitude; 8° Eastward from Zavijava, is η, of the 3rd magnitude, and nearly the same distance North of the Ecliptic. These five — β, η, δ, and ϵ Virginis, and β Leonis — form nearly a square, with the Southeast corner broken on the line joining η and δ. Two stars — o and ν — of the 4th magnitude, lying obliquely across the line joining the two β's, are in the head of Virgo, with π and ξ, of the 5th magnitude, near them. South of East from Zavijava, 27°, is α — a clouded white star of the 1st magnitude, named Arista, and sometimes called "Spica," or "the Spike of the Virgin;" it lies just South of the Ecliptic, 35° from Denebola, and 50° almost due South from Cor Caroli; the distance from α to β Virginis, is nearly the same as that from β Leonis to Cor Caroli. No. 61, of the 4th magnitude, 7¼° below Arista, marks the tip of the Southern wing; and θ, of the 4th magnitude, nearly on a line between Arista and δ, gives the location of the hand. A line from β, passing a little South of δ, and produced 17° farther, will touch τ, and 11° still farther is No. 109, both of the 4th magnitude, in the North Skirt; ζ, nearly on the Equinoctial, between δ and τ, is 11° above α. South 7½° from No. 109, crossing the Equinoctial, is μ, of the 4th magnitude, in the North foot, and 10° Southwest from μ is λ, of the 4th magnitude, in the right foot, near the Ecliptic, and 14° from Arista. The Virgin was supposed to represent the season of harvest.

90. East of Denebola 36°, Southeast of Cor Caroli, 26°, and nearly 35° from Arista, on the same line of Longitude, is Arcturus — a ruddy yellow star of the 1st magnitude — α, in Boötes. The Ecliptic is perpendicular to the line joining Arista and Arcturus. These two, with Denebola and Cor Caroli, form what is sometimes called the Diamond of Virgo. The three first named form nearly an equilateral triangle, while Cor Caroli is at the right angle of a triangle with Arcturus and

Map Nº VII.

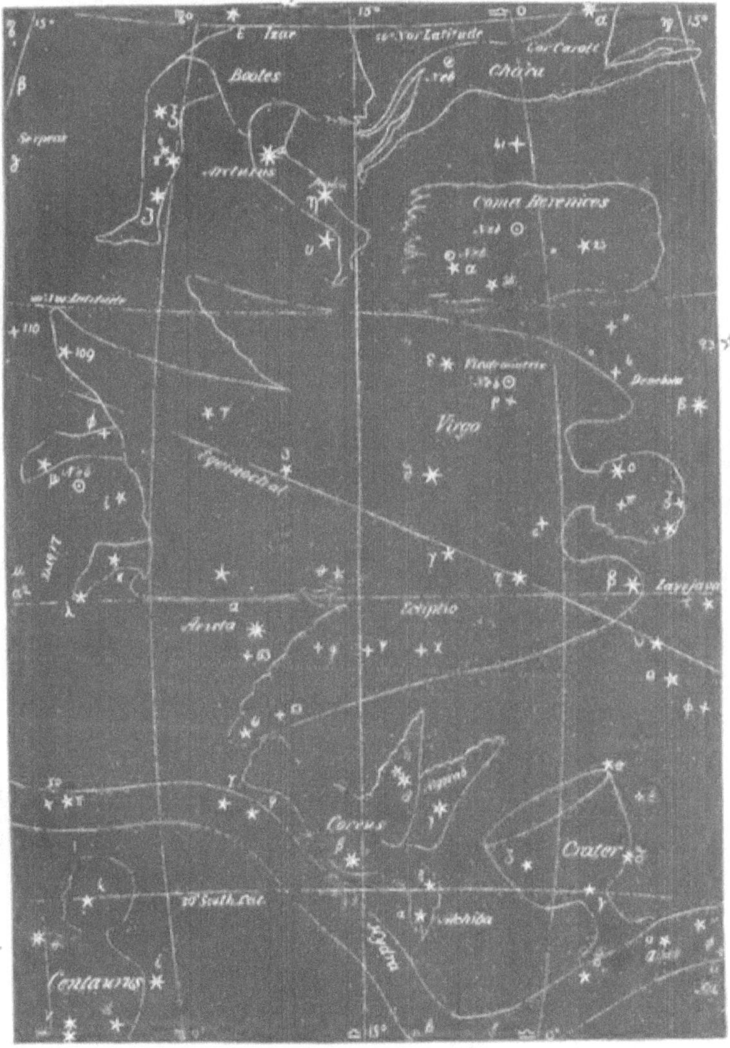

Izar
Bootes
Arcturus
Coma Berenices
Virgo
Equinoctial
Ecliptic
Corvus
Crater
Hydra
Centaurus

Denebola. From Arcturus, 5½° towards Denebola, is τ Bootis, of the 3rd magnitude, and 8° on the other side of Arcturus, is ε, also of the 3rd magnitude; the star outside the map — β Serpentis — is 22° beyond Arcturus, in the same line. ε is the most Northerly of a line of three stars, in the left leg of Bootes, all of the 3rd magnitude; the line is 6° in length towards the feet of Virgo.

91. COMA BERENICES — The Hair of Berenice. This constellation lies north of Virgo, between it and Canes Venatici. It contains 43 discernible stars, but only 3 of the 4th magnitude. α and No. 36 are easily found, being on a line midway between Denebola and Arcturus, and about 3° asunder. No. 23 is West and a little North from No. 36, forming with it, and α, an obtuse triangle.

92. CRATER — The Cup. This constellation lies Southwest of Virgo, and South of the tail of Leo, and contains 21 discernible stars. The line in the hinder leg of Leo (Sec. 86), continued 7° from ε Leonis, will touch θ in the Crater, and 8° farther is ζ, both of the 4th magnitude. From χ Virginis, through θ Crateris, and 6° farther South, is δ, of the 3rd magnitude; and 6° still farther is Alkes — α — an orange colored star of the 4th magnitude. Southeast 5° from α, is β, of the 4th magnitude, perpendicular to the line from α to θ; γ, a bright white star of the 4th magnitude, equidistant from α and β, is in a line from β towards θ.

93. CORVUS — The Crow. This constellation is East of Crater, and South of Virgo. It contains 9 stars visible to the naked eye; 6 of these are easily recognized as forming a peculiar diamond figure of 4° to 6° on a side, with a prominent star at the Southeast and Northwest angles, and a pair of smaller stars at each of the other angles. Alchiba — α — of the 4th magnitude, in the beak, the most Southerly star in that corner, is on the Equinoctial colure, 27° from Zavijava, almost in direct line with Denebola, and 23° Southwest from Arista; β, of the 2nd magnitude, is 6° East of α, towards the feet of Virgo.

94. HYDRA — The Snake. This constellation lies South of the Ecliptic, winding irregularly from below Cancer to opposite Libra; it is about 100° in length, and contains 60 discernible stars. Crater and Corvus rest on the body, near the middle. The arc of four stars in the head, with Alphard in the first fold, are shown on Map No. V (Sec. 78). From Alphard to Alkes — α Crateris — is 24°, and the forward or Western part of the body of Hydra is almost a straight line between these two points; μ, of the 4th magnitude, is 16° from Alkes, towards Alphard, and 6° farther West, almost equidistant from μ and α Hydræ, is λ, of the 4th magnitude, the only star in Hydra of any consequence not shown on the maps. From Alkes, some distance East, the position of the body is best traced by reference to the stars in Crater and Corvus. Just under the tail of the latter, 8° and 10° East from β Corvi, in a line with ε, and just below there the end of the Southern wing of Virgo, are ξ and γ, both of the 4th magnitude; 11° from γ, and 18° below Arista, is π, of the 4th magnitude, in a line with Arista and δ Virginis.

95. A line from Cor Caroli through Arista, and produced 23° farther, will pass through the head of the Centaur, the upper part of which constellation contains a group of stars similar in arrangement to that in Corvus, but composed of larger stars. ε, of the 3rd magnitude, in the shoulder, is 17° Southeast from β Corvi, nearly in a line with γ Corvi.

DEFINE AND EXPLAIN (the figures refer to the sections):

88. Map No. 7; Connections. 89. Virgo: Zavijava; Vindemiatrix; broken square; Arista; north Skirt. 90. Arcturus; diamond of Virgo; left leg of Bootes; 91. Coma Berenices; obtuse triangle. 92. Crater; Alkes. 93. Corvus; diamond; Alchiba. 94. Hydra; Alphard to Alkes. 95. Head of Centaur.

MAP NO. VIII.

96. This map comprises a section of the heavens next East of that shown in the map preceding, the Ecliptic running nearly through the middle of the map; but, as this part of the Ecliptic is far south of the Equinoctial, the stars shown near the lower margin are never seen far above the horizon of the Northern States. The map shows the constellations Libra, Scorpio, Ophiuchus, and Lupus; with portions of Hydra, Centaurus, Telescopium, Taurus Poniatowski, and Ara. It includes, in common with the preceding map, δ and β in the head of the Serpent, μ and λ in the feet of Virgo, a^2 Libræ, π Hydræ, and ι, μ, $\varkappa$, and θ Centauri, with a few smaller stars. The Right Ascension of the center of the map is about 16 hours, and it is on the Meridian that number of hours after the Dipper is below the Pole (See. 36). The Stars in the upper left hand corner are very near the 15 hour circle, which runs through the beginning of $\mathcal{V}_{2}$.

97. Libra — The Balance. This is the eighth in order of the Zodiacal constellations; it lies East of Virgo, South of the Equinoctial, and on the Ecliptic, the greater number of its stars lying between those two circles. It contains 51 discernible stars. Zubenesch — a^2 — a pale yellow star of the 3rd magnitude, lies just North of the Ecliptic, 8° from λ, in the South foot of Virgo, 21° East from Arista, and 11° South from μ Virginis, forming with λ and μ, a nearly isosceles triangle, whose base is half the height. East from μ Virginis 9°, in line from Vindemiatrix, is Zubenelg — β — a pale emerald colored star of the 2nd magnitude; a^2 and β form with μ and λ Virginis, a nearly diamond trapezium;

and δ, of the 4th magnitude, 4° West from β, towards Arista, forms a smaller trapezium with the other three. No. 37, of the 4th magnitude, is about as far East from β; and No. 51, of the 4th magnitude, in the Eastern edge of the constellation, is 7½° still further East, being 12° from β. Southwest by South from No. 51, are two stars, of the 4th magnitude — θ, 6° below; and from θ two other stars of the 4th magnitude — γ, and γ — run West nearly 5° towards δ, while ζ, of the 4th magnitude, is 2½° Southwest, and 10° farther, in the same line, is No. 20, of the 3rd magnitude, in the Southern limit of the Scales. A few stars of the 5th magnitude, lying near these, will be easily located.

The step-like arrangement of Nos. 51 and 41, and θ, ζ, γ, ζ, and No. 20, is readily recognized in the heavens, though the stars are only of the 4th magnitude. μ Serpentis forms, with Zubenelg and Zubenesch, a line of equidistant stars stretching from 2° South of the Equinoctial to the Ecliptic. The constellation Libra was so named because when the Sun formerly entered that group of stars the days and nights were equally *balanced*. Libra is the most modern of the ancient constellations; on the earlier Zodiacs the claws of the Scorpion were extended to the feet of Virgo.

98. Scorpio — The Scorpion. This is the ninth of the Zodiacal constellations; it is situated Southeast from Libra, and contains 44 discernible stars. It is recognized from the peculiar arrangement of the stars in the Southeastern portion. The principal star, Antares — a — a fiery red star of the 1st magnitude, is 46° Southeast by East from

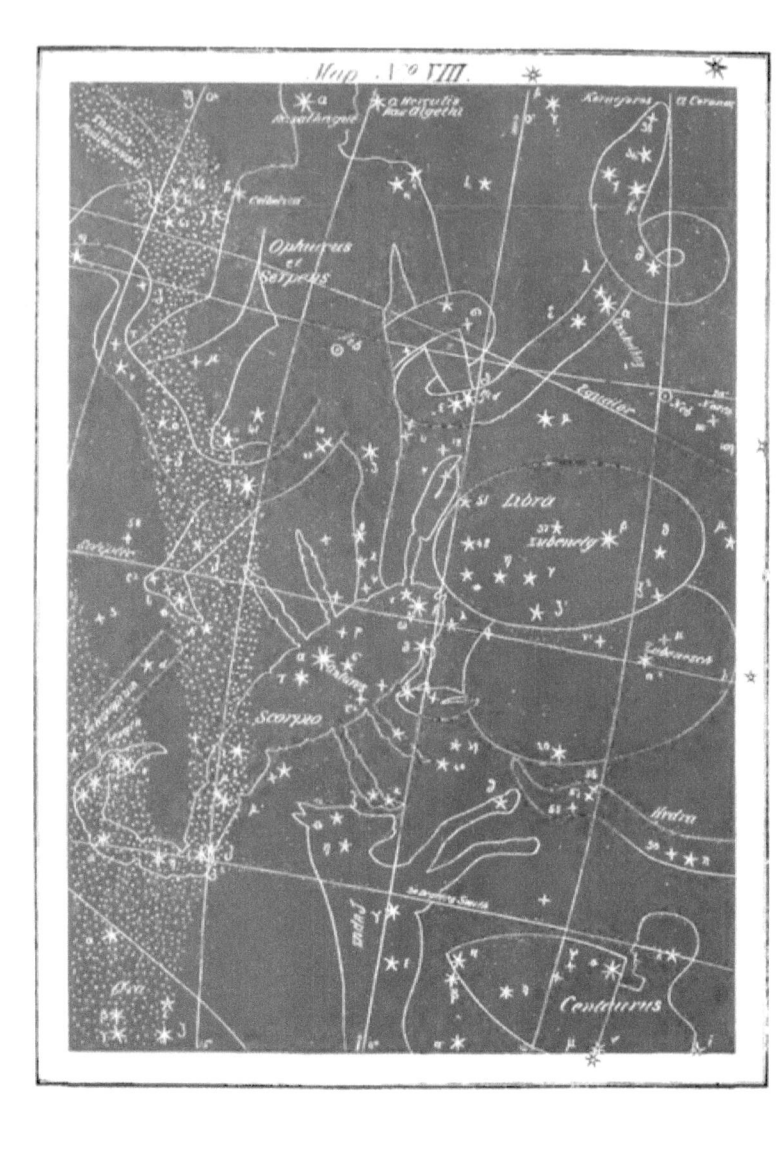

Arista, and 56° Southeast from Arcturus. It is located near the middle of the body, and hence is often called "Cor Scorpio"—the heart of the Scorpion.

Nearly equidistant from α, in line with Arcturus, are Graffias—β—a pale white star of the 2nd magnitude, above, and ε of the 3rd magnitude, below α; 2° on each side of α are σ, of the 4th magnitude, and τ, of the 3rd; two smaller stars also lie near β. From these, ε, ρ¹, ρ², ζ¹, ζ², τ, θ, ι, κ, υ, and λ, run in the order named, forming the tail, and describing nearly three-fourths of an ellipse. They are all of the 3rd magnitude; Lesath — λ — in the extremity of the tail, is 19° Southeast of Antares in line with Zubenelg. The claws contain many stars of the 4th magnitude; ξ, ρ, π, and δ, forming a line of 12° in length, running due South from the head. The position of ψ, of the 5th magnitude, in the upper forward claw, is best noted by remembering that it is 3° East from No. 51 Libræ — the uppermost star in the step of Libra. Scorpio was probably so named to memorize the fact that when the Sun was among these stars, the diseases of Autumn were prevalent. The positions of the stars are, however, naturally suggestive of the shape chosen.

99. The tail end of Hydra is shown below Libra; it contains no prominent star not already noted. Farther South are the head and upper part of the body of Centaurus (See. 95). The two stars of the 3rd magnitude, at the Eastern end of the figure, are α Centauri and β Lupi; they are about 1° apart, and 26° and 27° South of Zubenesch, 37° Southeast of Arista, and 24° Southwest of Antares. Between Centaurus and Scorpio is Lupus — the Wolf. (See Sec. 134.) Northeast of the tail of Scorpio is Telescopium. (See Sec. 113.)

100. ARA — The Altar. This constellation lies South of the tail of Scorpio, and contains 9 discernible stars. α Aræ is 13° South of Lesath; β and γ are in the same line 6° and 7° farther South, and ζ and ε lie West of β and γ; all these, except ε, are of the 3rd magnitude.

101. TRIANGULUM AUSTRALIS — Three stars, 7°, 8° and 6° apart.—α, of the 2nd magnitude, and β and γ, of the 3rd, form the Southern Triangle. It lies 10° Southwest by South of β Aræ, and 8° East of α Centauri. (See Maps XII. and XIII.) The constellation contains 5 discernible stars.

102. OPHIUCUS ET SERPENS — The Serpent and his Bearer. This constellation is North of Libra and Scorpio. It occupies a large space in the heavens, and contains 138 discernible stars, of which 74 are catalogued as in Ophiucus, and 64 as in Serpens. β — a pale blue star of the 3rd magnitude, in the head of the Serpent, is 31° East of Arcturus, and with γ, of the 3rd magnitude, 2° East, and No. 34, of the 4th magnitude, above, forms a small triangle in the head; with No. 38, of the 5th magnitude, they mark the location of the head. Unukalhay — α Serpentis — a pale yellow star of the 2nd magnitude, is 9° South from β, 25° Southeast by East from Arcturus, and 34° Northwest by North from Antares; δ — a bright white star of the 3rd magnitude in the first coil — is 4° above α, towards the upper part of Bootes; and β° below α is ε, of the 3rd magnitude, with λ, of the 4th, just East of α. Southeast from α, are δ — Yed — and ε Ophiuci, both of the 3rd magnitude, and 1° apart, in the right hand of Ophiucus, in line with α and ε Serpentis. The Serpent now winds back around the arm, taking in one star of the 4th magnitude, and two of the 5th, and then stretches Southeast across the body, taking in η Ophiuci, of the 2nd magnitude, 17° beyond ε, in line with α; then winds up Northeast, including ω, ν and χ, of the 4th magnitude, 5° and 9° apart. The tail extends some distance farther than the limits of the map, but only contains one other noteworthy star — θ, of the 4th magnitude, about 11° beyond ζ. East of β 27°, nearly on a line with Arcturus, and 28° a little North of East from Unukalhay, is Rasalague — α Ophiuci — in the head, a Sapphire colored star of the 2nd magnitude; 8° below α Ophiuci is Celbelrai — β — of the 2nd magnitude, in the left shoulder, in line towards τ Serpentis, with γ, of the 4th magnitude, about 2° farther in the same line. The right shoulder is marked by κ and ι, both of the 4th magnitude, 13° from β towards the head of Serpens, and with β forming a triangle with Rasalague. ζ, of the 3rd magnitude, in the right knee, is 21° below κ and ι, towards Antares. The lower part of the right leg is marked by a line of three small stars below ζ, extending nearly to θ Scorpii, the lowest — θ — being 10° from ζ. The left leg contains η, of the 4th magnitude, 6° below χ, and 4° farther are three stars in the foot, including θ, of the 3rd magnitude, 12° East of Antares.

103. The group of stars 5° Southeast of the left shoulder of Ophiucus, are in the face of Taurus Poniatowski. West of Rasalague 5°, is Ras Algethi — α — an orange colored star of the 3rd magnitude in the head of Hercules; and 14° farther West are β and γ Herculis, two silver white stars of the 3rd magnitude, 3° apart, the line passing through them pointing to Unukalhay. East and a little North from Arcturus, 20°, is Alphecca — α — Coronæ Borealis — a brilliant white star of the 2nd magnitude, 24° North of Unukalhay, and 14° from β Herculis, in line with Rasalague.

South of the body of Scorpio, between Ara and Lupus, and North of Triangulum Australis, is a space about 10° broad, and 25° in length from North to South, which has been filled in by modern astronomers with the figure of a squaring tool and ruler, and called Norma Euclidis — the Square of Euclid — in honor of the great geometrician. It contains no important stars. Immediately South of Lupus, and West of Triangulum, is another small geometrical constellation called "Circinus" — the compasses. This constellation contains but one important star — α, of the 4th magnitude, and shown on Map XIII., at the feet of Centaurus.

The Milky Way skirts the region represented near the left margin of the map; through the head of Taurus Poniatowski in a narrow stream, and takes in the left shoulder and arm of Ophiucus to the first two joints in the tail of Scorpio, there uniting at the bend of the tail with another stream which includes the extremity of the tail of Scorpio, and flowing through Ara in one broad stream; then North of Triangulum, and through Circinus.

DEFINE AND EXPLAIN (the figures refer to the sections):

96. Map No. VIII.; connections. 97. Libra; Zubenesch; Zubenelg; diamond; the step. 98. Scorpio; Antares; Graffios; course to Leath; upper forward claw. 99. Centaurus and Lupus. 100. Ara. 101. Triangulum Australis. 102. Ophiucus et Serpens; triangle in head; Unukalhay; Yed; Eta; Rasalague; Celbelrai; right shoulder; the legs. 103. Taurus Poniatowski; Alphecca; Norma Euclidis; Milky Way.

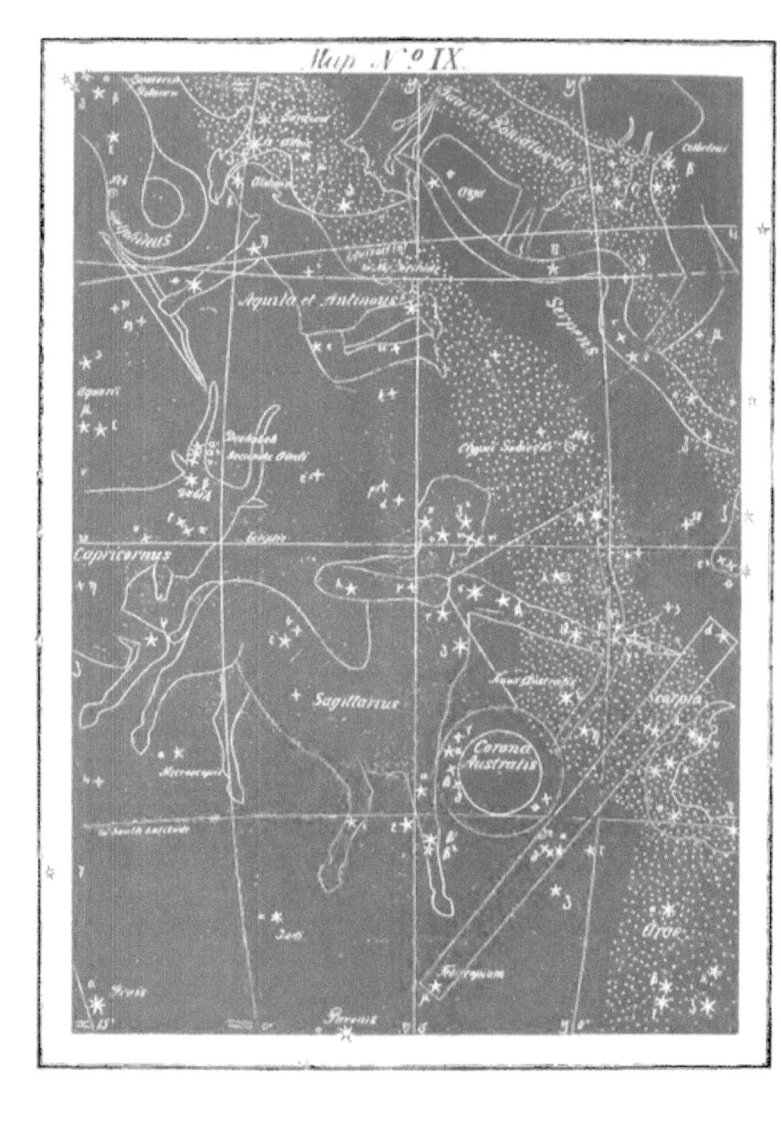

MAP NO. IX.

104. This map represents a section of the heavens next East of that shown in the preceding map; its center has a Right Ascension of about 19 hours, and comes to the meridian about 5 hours before the Vernal Equinox. It contains, in common with Map No. VIII., the left shoulder and head of Ophiucus, the tail end of Serpens, the tail of Scorpio, and the principal stars in Ara.

105. TAURUS PONIATOWSKI — The Bull of Poniatowski — is a small constellation, containing only 5 noteworthy stars, located East of the left shoulder of Ophiucus, and catalogued as belonging to that constellation. Three of those stars are of the 4th magnitude, and form a small triangle 5° Southeast of Celbelrai.

106. AQUILA ET ANTINOUS — The Eagle and Antinous. This constellation lies on the Equinoctial, East of the body of Ophiucus, and contains 74 discernible stars. East 32° from Celbelrai, 31° from Rasalague, and 8° North of the Equinoctial, is Altair — α — a pale yellow star of the 1st magnitude, in the neck of the Eagle, and the middle one of three prominent stars; γ, 2° above, and β, 3° below, both pale orange stars of the 3rd magnitude; 8° farther Southeast by South, in the line of these, is θ, of the 3rd magnitude, in the hand of Antinous; Southwest from Altair 9°, and 12° from θ, is δ, of the 3rd magnitude, making a triangle with α and θ, also with β and γ; 9° below δ, in a line from γ, is λ, of the 3rd magnitude, in the heel of Antinous; ζ, of the 4th magnitude, is in line between θ and δ, the Equinoctial passing nearly midway between ζ and θ. The two stars of the 5th magnitude, Southwest of λ, belong to an obscure constellation called "Clypei Sobieski" — the shield of Sobieski — not outlined on the map.

107. DELPHINUS — The Dolphin. This constellation lies East of Aquila, and contains 18 discernible stars. Nearly Northeast 11° from Altair, is Svalocin — α — a pale white star of the 3rd magnitude, situated at the Northern side of a small diamond figure of stars, about 2° apart. Rotanen — γ — a green tinted star of the 4th magnitude, is at the Southwest corner of the figure; 5° South from Svalocin, is ε, on a line drawn through Rotanen.

108. SAGITTARIUS — The Archer. This is the tenth in order of the Zodiacal constellations, lying East of Scorpio, South of the Ecliptic, and on the Solstitial colure, or 18 hour circle; it contains 69 discernible stars, and is represented on the globe as a Centaur — half man and half horse — with bow and arrows, his head on the Ecliptic. The most prominent stars are four of the 3rd magnitude — δ, ε, ζ, and σ, forming a rhomboid, or elongated diamond figure, about twice as long as broad; δ, in the hand, is 25° due East from Antares, and 9½° beyond, in the same line, is ζ, in the breast, the line joining them being the shortest diagonal of the rhomboid; ε is 5° below δ, and about 10° from the end of the tail of Scorpio; σ, in the forward shoulder, is 5° above ζ. North 9° from δ, 11° Northwest from σ, and 16° Southeast from ζ in the left knee of Ophiucus, is μ, the last of a slightly curved line of nearly equidistant stars, composed of Unukalhay, Yed, ζ Ophiuci, and μ Sagittarii. γ¹ and γ², both of the 4th magnitude, nearly 3° West of δ, are in the crown of the bow.

The place of the head of Sagittarius is marked by three stars of the 4th magnitude, 5° above σ, and just North of the Ecliptic; α, of the 4th magnitude, is 12° nearly South from ζ, on a line with σ; 5° below α are β¹ and β², of the 4th magnitude, in the forward leg. The only remaining star of note in the constellation is ε, of the 4th magnitude, 11° East from ζ, in line with δ. The constellation Sagittarius was probably so named because the Sun was formerly passing through it at the season for hunting in the East.

109. CORONA AUSTRALIS.—The Southern Crown. Northwest 1° from α Sagittarii, is α Coronae Australis, of the 4th magnitude, in a curved line of four stars on the Eastern side of the constellation.

110. MICROSCOPIUM.—The Microscope. Southeast 12° from ε Sagittarii is α Microscopii, of the 4th magnitude, nearly between the hind feet of Sagittarius; it is the only star of note in the constellation.

111. GRUS.—The Crane. Southeast 14° from α Microscopii, in line with ε Sagittarii, is γ, of the 3rd magnitude, in Grus; 10° South from γ is α Gruis, of the 2nd magnitude; β, of the 3rd magnitude, is 6° East from α.

112. INDUS ET PAVO.—The Indian and Peacock. South 14° from α Microscopii, is α Indi, of the 3rd magnitude, forming with it and α and γ Gruis, a nearly square figure. South 10° from α Indi, is α Pavonis, of the 2nd magnitude, in the eye of the Peacock; and 10° South from α Pavonis, is β Pavonis, of the 3rd magnitude—the middle one of a line—δ of the 4th magnitude, and β and γ of the 3rd magnitude; the line is 8° in length, and runs East and West on the Antarctic circle;

therefore not visible in the United States. α Pavonis is 25° East from β and γ Arae, and 35° Southeast from Lesath. α Pavonis, α Indi, and α and γ Gruis, form a large diamond figure. Indus et Pavo is a double constellation, like Aquila et Antinous, and extends nearly to the South Pole.

113. TELESCOPIUM.—The (Southern) Telescope. Between α Pavonis and λ Scorpii, and between Sagittarius and Ara, is a group of stars in the middle of the Southern Telescope—α, ε, and ζ—all of the 4th magnitude, and forming a small triangle, α forming also a smaller triangle with α¹ and α², of the 5th magnitude. μ Telescopii, of the 4th magnitude, is 14° below α, towards α Pavonis; and γ, of the 4th magnitude, is 3° East of Lesath, and 12° above α; β is 6° East of γ.

114. A little East of South, 25° from Altair, and 23° a little North of East from ε Sagittarii, in line with δ, is Dabih—β—an orange tinted star of the 3rd magnitude, in the head of Capricorn; 2½° above Dabih is Secunda Giedi—α²—a pale yellow star of the 3rd magnitude, and just above α², is Deshabeh—α¹—of the 4th magnitude, also called "Algedi;" 12° below β, and 10° East from ε Sagittarii, is ζ, of the 4th magnitude, in the knee of Capricorn.

The main stream of the Milky Way comes within the limits of the map at the lower right hand corner, including Ara and the tail of Scorpio. At the tail it divides into two narrower streams; the Eastern one takes in the bow of Sagittarius, shield of Sobieski, the feet of Antinous, and skirting his back, bends to include the whole of Aquila; the Western stream includes the left foot, hand, and shoulder of Ophiucus, with a part of the tail of Serpens, and Taurus Poniatowski.

DEFINE AND EXPLAIN (the figures refer to the sections):

104. Map No. IX; connections. 105. Poniatowski's Bull. 106. Aquila et Antinous; Altair; Clypei Sobieski. 107. Delphinus, Sobieski; Rotanen. 108. Sagittarius; Rhombus; Unukalhay to Mu; the bow. 109. Corona Australis. 110. Microscopium. 111. Grus. 112. Indus et Pavo. 113. Telescopium. 114. Secunda Giedi; Algedi; Milky Way.

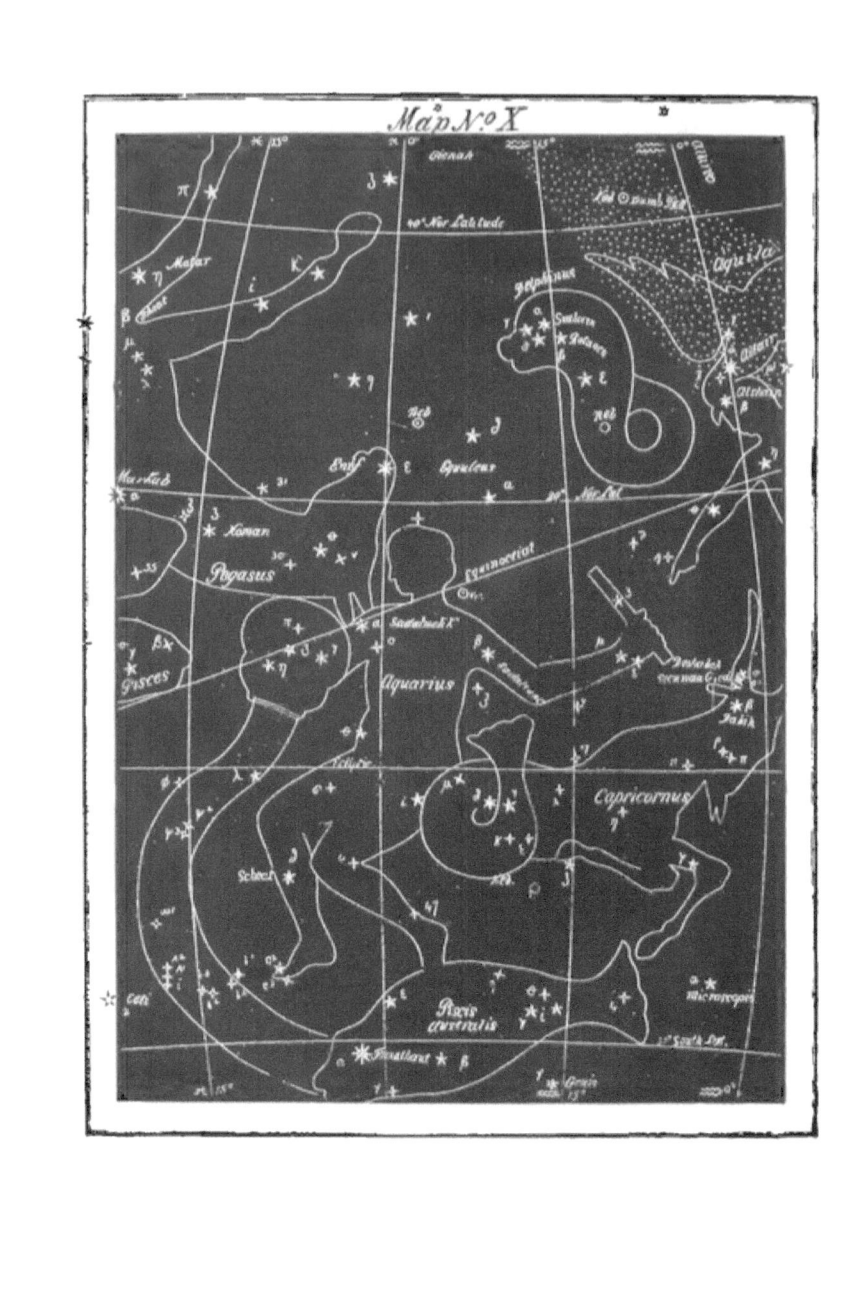

MAP NO. X.

115. This map represents a section of the heavens next East of that shown in the map preceding; the Right Ascension of its center is about 22 hours, and it comes to the meridian above the earth 2 hours before the Vernal Equinox, and a little more than half way from the Zenith towards the Southern horizon. It contains, in common with Map No. IX., the heads of Aquila et Antinous, Delphinus, the head of Capricorn, stars in the hand of Aquarius, α Microscopii, and γ Gruis.

116. CAPRICORNUS — The Goat. This is the eleventh in order of the Zodiacal constellations, and contains 51 discernible stars. It lies on the Ecliptic, East of Sagittarius, and is represented on the globe as a goat with the tail of a fish. The principal stars are those already named (Sec. 114) — β in the head, and a^1 and a^2, in the left horn. Below β. 4°, are π and ρ, both of the 5th magnitude, in the lower part of the face; 21° East from β is δ, of the 3rd magnitude, in the tail, and 2° West of δ is γ, of the 4th magnitude; ζ, of the 4th, is 8° South from δ, and 10° a little North of East from ψ in the knee (See. 114); χ and ε, of the 5th magnitude, between δ and ζ, and μ, of the 5th, a little North of δ, form a line marking the position of the tail, and its junction with the next constellation; the line points towards α Microscopii. Capricornus was probably so named to signify that when the Sun was among those stars he began to move upwards, towards the North, the goat being a climbing animal. The fish tail was a later mythological addition.

117. AQUARIUS — The Water Bearer. This is the twelfth in order of the Zodiacal constellations; it lies on the Ecliptic, East of Capricorn, West of Pisces, and contains 108 discernible stars. It is represented as a man in nearly a sitting posture, with an inverted urn in his grasp, from which flows a stream of water. The two principal stars are both of the 3rd magnitude, and a pale yellow color, one in each shoulder. Sadalsund — β — is 20° a little North of East from Dabih; and Sadalmelik — α — is 10° farther in the same line, also 21° Southwest from Markab, in the Square of Pegasus (See. 56), nearly on a line with Alpheratz, and about as far distant from Markab as is that star; α is also 29° East from θ Aquilæ, and, with it, marks very nearly the course of the Equinoctial. Sadalsund is 25° Southeast from Svalocin. Midway between Deshabeh and Sadalsund are μ and ε, of the 4th magnitude, in the hand, and 4° North from these is No. 3, of the 4th magnitude, nearly on the line between Sadalsund and θ Aquilæ; 20° Southeast from α, and 16° East from δ Capricorni, is Scheat — δ — of the 3rd magnitude, in the forward knee; θ and ι, of the 4th magnitude, 7° apart, lie nearly midway across the line from β to δ; 7° a little East of South from δ is c^2, of the 4th, with c^3, of the 5th magnitude, marking the position of the forward foot.

East 6° from α is ζ, of the 4th magnitude, the central star in a Y-shaped figure formed by γ, of the 3rd, ζ, η, of the 4th, and π, of the 5th magnitude — the top towards α — and marking the place of the Urn. The course of the Stream is marked by a number of small stars almost in a semicircle round the knee and feet of the figure; λ — the only star of the 4th magnitude in the stream —

is 10° from γ in the Urn, and 20° East from β. East 13° from c² is No. 2 Ceti, in the tail of Cetus (Sec. 62). East 12° from the middle of the Y is γ Piscium, of the 4th magnitude, in the head of the Western Fish. (See. 55.) The constellation Aquarius, probably received its name from the fact that the rainy season was coincident with the course of the Sun among those stars.

118. PISCIS AUSTRALIS — The Southern Fish. This constellation is South of Aquarius, and West of Cetus, and contains 24 discernible stars. It is represented with open mouth, drinking the water in the stream of Aquarius. Fomalhaut — α — a reddish star of the 1st magnitude, in the mouth, is 27° Southwest from Diphda in Cetus, 20° Northeast from α Gruis, 16° nearly East from γ Gruis, 38° Northeast from α Pavonis, 22° Northwest from α Phœnicis, 45° due South from Markab, in the West side of the Square of Pegasus, 32° Southeast from β Aquarii, 39° a little South of East from β Capricorni, 21° Southeast from δ Capricorni, 40° Northwest from Achernar — α Eridani — 60° Southeast from Altair, and 27° East from α Microscopii. It forms a nearly equilateral triangle with ε, 5°, and β, 6°, distant, both of the 4th magnitude, β being nearly midway between Fomalhaut and α Gruis. West 10° from β, 16° West from α, and 5° above γ Gruis, is ι, of the 4th magnitude. 9° West of ι is η, of the 5th magnitude. α, ε, η, and θ, form an arc of stars outlining the back of the fish.

119. PEGASUS — The Winged Horse. This constellation is North of Aquarius and Piscis Occidens, Southwest of Andromeda, and East of Delphinus; it contains 89 discernible stars. It is represented on the globe as the forward part of a flying horse. Algenib — the Southernmost star in the Eastern side of the Square — (Sec. 50) is the only one, of note in the constellation, not shown on this map. Enif — ε — of the 2nd magnitude, in the nose, is 20° a little South of West from Markab, 12° above Sadalmelik, 16° above Sadalsund, 28° East from Altair, and 17° a little South of East from Svalocin.

The junction of the head and neck of Pegasus is marked by three small stars, midway between Enif and the Y of Aquarius; and between these and Markab, 7° from the latter, in line with Alpheratz, is Homan — ζ — of the 3rd magnitude. North 19° from ζ, and 5° West from β, is Matar — η — of the 3rd magnitude; 6° from η, and 5° from β is π, and near it is λ, both of the 4th magnitude, in the breast; π and ι, both of the 4th magnitude, in the legs, form, with η, a nearly equilateral triangle of 8° on each side; χ, of the 4th magnitude, is 5° West from ι.

120. EQUULEUS. The Little Horse. The two stars of the 4th magnitude, 5° apart, and 8° to the West from Enif, forming with it a triangle, belong to Equuleus — a constellation containing 10 discernible stars; the head only is shown on the globe, in front of the head of Pegasus.

121. The star marked ζ, 20° West from Matar, also Gienah and Albireo — the two large stars in the upper margin of the map — are in Cygnus. The course of the Milky Way is shown in the upper right corner. The Eastern stream includes Aquila (Sec. 114), and runs Northeast towards Cygnus.

DEFINE AND EXPLAIN (the figures refer to the sections):

115 Map No. X; connections. 116 Capricornus; the head and horn; the tail. 117. Aquarius; Sadalsund; Sadalmelik; Scheat Aquarii; the Y; the stream; 118. Piscis Australis; Fomalhaut; Equilateral triangle. 119 Pegasus; the Square; Enif, Homan; Matar. the rest. 120. Equuleus 121. Cygnus; the Milky Way.

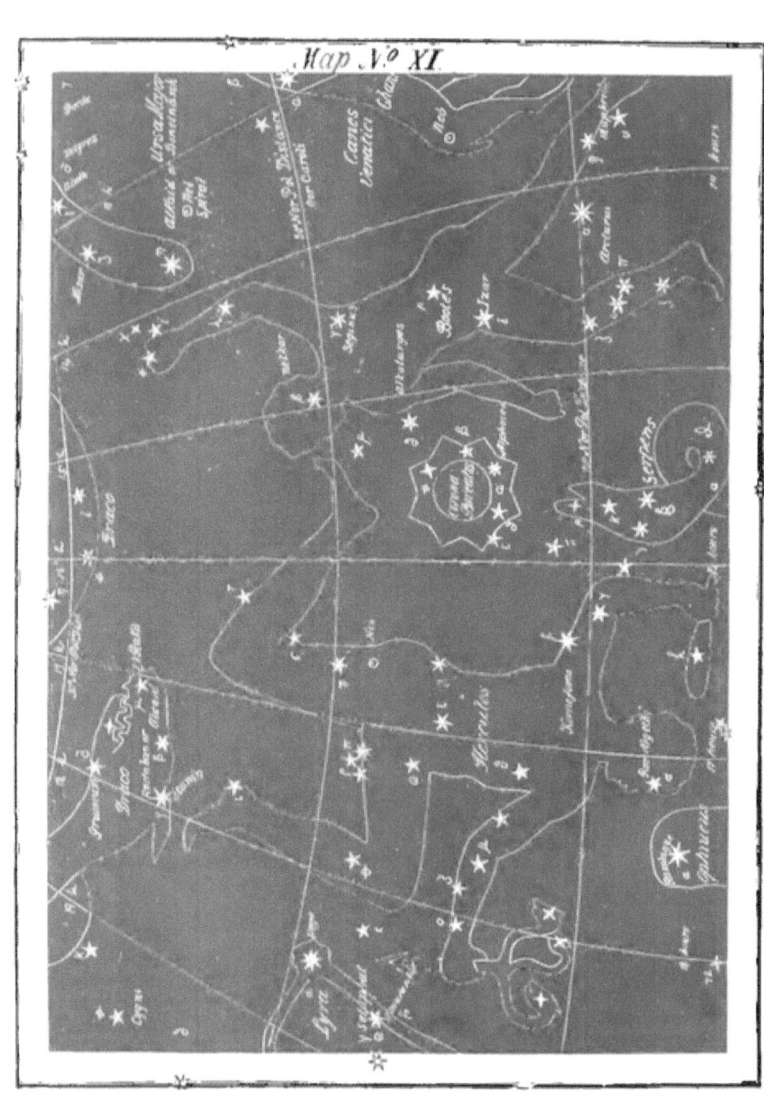

MAP NO. XI.

122. This map is projected with the Equinoctial as a base line, instead of the Ecliptic. It represents that portion of the heavens situated between Draco and Ophiucus et Serpens. The perpendicular through the center of the map is a portion of the 16 hour circle of Right Ascension; it is on the meridian 2 hours before the head of Draco (Sec. 44), and 16 hours after, or 8 hours before, the Vernal Equinox. The map contains the handle of the Dipper and head of Draco, in common with Map No. I., Cor Caroli and the lower part of Bootes, in common with Map No. VII., and the heads of Ophiucus and Serpens, in common with Map No. VIII.

123. BOOTES. The Bear Driver. This constellation is represented on the globe as a man holding in his hand a leash, directing two dogs — Asterion and Chara — which are hunting the Bears round the Pole. It contains 54 discernible stars. The principal star is Arcturus — a — a reddish yellow star of the 1st magnitude, situated (See. 96) 33° Northeast from Arista, 26° Southeast from Cor Caroli, 31° nearly South from Alkaid, 26° Northwest from Unukulhay, 40° West of Rasalague, and 54° Southwest from Etamin. East from Arcturus 8°, is ε, of the 3rd magnitude, in the left knee; 6° below ε is ζ, of the 3rd magnitude, which is also 8° Southeast from a, these three forming an isosceles triangle; π and u, of the 3rd magnitude, are nearly equidistant from ε and ζ, a little nearer a, and on the line from a to Unukulhay. West 6° from a, nearly on a line with ε, is η — Muphrid — of the 3rd magnitude; 10° Northeast from a is Izar — ϵ — and 9° farther in the same line is Alkaturgos — δ — a golden yellow, in the left shoulder; ϵ is 9° above ε, and 12° farther is Seginus — γ — a silver white star in the right shoulder; ϵ, δ and γ are all of the 3rd magnitude; a, γ, δ and ε, form a four sided figure, with ϵ at the intersection of the two diagonals; and ρ, of the 4th magnitude, is 4° above ϵ, in line between a and γ. Nekkar — β — a golden yellow star of the 3rd magnitude in the head, is 8° above δ, 6° north of east from γ, and 24° above a. From β, 10° towards Alkaid, is λ, of the 4th magnitude, and 5° north of λ, and about as far from Alkaid, is ι, of the 4th magnitude; ι forms a small right angled triangle with θ, of the 4th, and χ, of the 5th magnitudes, in the hand of Bootes.

124. CORONA BOREALIS—The Northern Crown. This constellation is situated east of Bootes, and contains 21 discernible stars. Southeast 8° from Alkaturgos, towards the head of Serpens, is Alphecca — a — a brilliant white star of the 2nd magnitude, also 21° North from Unukulhay, and 20° Northeast from Arcturus — the three forming a prominent triangle of nearly equal sides. From a, 3° towards Alkaturgos, is β, of the 4th magnitude. These two form a semi-circle with ϵ, δ, and θ. The other stars in this constellation are small.

125. HERCULES — The Sampson of Greece. This constellation is East of Bootes, North of Ophiucus, and South of Draco, contains 113 discernible stars, and is represented on the globe as a man in an inverted position, kneeling, holding a club in one hand, and the three-headed Cerberus in the other. The principal star is Kornetoros — β —a pale yellow star of the 2nd magnitude in the

West shoulder, situated 32° East from Arcturus, 35° Southwest from Etanin, and 44° Southeast by East from Alphecca, in line with Rasalague. Southeast from β, 13°, is Ras Algethi — α — an orange colored star of the 3rd magnitude in the head, nearly on a line towards Rasalague, and 5° from that star. Southwest 3° from β, towards Unukulhay, is γ — a silver white star of the 3rd magnitude; 12° on the other side of β, in the same line, is ϵ, of the third magnitude, 7° farther is π, of the 3rd, and 14° still farther, in the same line, is ι, of the 4th magnitude, only 6° South of Etanin. These seven stars form a slightly curved line of 54° in length from Etanin to Unukulhay. β and γ, with χ, and with γ and β Serpentis, form another curved line of Stars.

East 2° from π is ρ, and 6° farther is θ, both of the 4th magnitude, these marking the position of the Eastern thigh, and forming a triangle with ι, of the 4th magnitude, in the foot. North 44° from α, Northeast 12° from β, and 7° below ϵ, is δ — a greenish tinted star, of the 4th magnitude. A line from δ, through ϵ, produced 9° farther, will nearly touch ζ, of the 3rd magnitude, also σ, 4° farther, and τ, 5° still farther — the last three marking the position of the Western thigh. A line from δ, through δ, will pass μ, ξ, and o — three equidistant stars, of the 4th magnitude, in the Eastern arm; 7° South of these are three stars of the 5th magnitude in the heads of Cerberus, 20° North of the head of Taurus Poniatowski.

126. LYRA — The Harp. This is a small constellation, containing 21 discernible stars, lying East of Hercules, and North of the head of Draco. A little East of South 15° from Etanin, 30° Northeast from Rasalague, 60° a little North of East from Arcturus, 32° Northeast from Korneforos, 35° a little West of North from Altair, and 40° a little North of East from Alphecca, is Vega — α — a sapphire colored star of the 1st magnitude; 6° farther from Etanin is Sheliak — β — a white star of the 3rd magnitude, and 2° Southeast from β is Sulaphat — γ — a bright yellow star of the 3rd magnitude.

The stars marked δ, θ and ϵ, in the upper left hand corner of the map, South of the first fold of Draco, belong to Cygnus. The Solstitial Colure, coincident with the 18 hour Circle, is represented as a curved line on the map. It passes near Grumium and Etanin, in the head of Draco (See. 44), West of Vega, through the heads of Cerberus, touches No. 72 Ophiuci, cuts the head of Taurus Poniatowski, and intersects the Ecliptic in the first point of $\vee$, West 1½° from μ Sagittarii (See. 108). The star No. 72 Ophiuci, in the Southeast corner of the map, lies in the Milky Way, which runs East of Cerberus and Lyra, from the left shoulder of Ophiucus.

DEFINE AND EXPLAIN (the figures refer to the sections): 122. Map No. XI.; projection; connections. 123. Boötes; Arcturus; isosceles triangle; Mupfrid; Alkaturgos; Izar; Nekkar. 124. Corona Borealis, Alphecca. 125. Hercules; Ras Algethi; Etanin to Unukulhay; Eastern arm; Cerberus. 126. Lyra; Vega; Sheliak; Sulaphat; Cygnus; Solstitial Colure; Milky Way.

Map N.º XII.

MAP NO. XII.

127. This map represents a portion of the heavens extending from the parallel of the Winter Solstice — 23½° of South Declination — to a few degrees past the South Pole; the perpendicular line is the circle of 8 hours of Right Ascension. The map includes the Southern portion of Canis Major, in common with Map No. V., the whole of the constellation Argo, and contiguous stars, many of which have been previously noted.

128. Argo — The ship in which Jason and his companions are said to have sailed in quest of the golden fleece of Aries. This constellation covers a large portion of the heavens lying Southeast of Canis Major, but the greater number of its stars are always below the horizon of the Northern States. It contains 64 discernible stars, two of which — ρ and ξ, both of the third magnitude — in the beak of the prow, are shown on Map No. V. Tureis — ρ — is 21° Southeast from Sirius, and 25° Southwest from Alphard; ξ is 4° West from ρ. South 37° from Sirius, 21° Southeast by South from Phact, and 30° East from Achernar, is Canopus — α — of the first magnitude, under the front of the keel.

Southeast 29° from Sirius, in line with Betelgueuse, 16° South from Tureis, and 21° Northeast from Canopus, is Naos — ζ — of the 2nd magnitude, in the forward deck. Southeast by South 25° from α, is β, of the 1st magnitude, under the stern, with 9h. 12m. of Right Ascension, and 69° South Declination, or 159° of North Polar Distance. Above β, 12° towards Aludra, and 18° Southeast by East from α, is ε, of the 2nd magnitude; East 7° from ε, and 25° from α, is ι, of the 2nd magnitude; East 11° from ε is η, of the 2nd magnitude; θ, of the 3rd magnitude, is 5° South from ι, and μ, of the 3rd, is 11° North from ι. These three — μ, ι, and θ — form a line North and South, with 10h. 40m. of Right Ascension, and mark the stern, or Eastern, line of the ship; on the globe they are generally represented as in the body of a tree called Robur Caroli, or Charles' Oak, but they are catalogued as in Argo, and the tree is not placed on these maps. North 15° from β, and 4° above ι, is κ, of the 3rd magnitude; these three forming another line running North and South, with a Right Ascension of about 9h. 15m. Southeast 7° and 12° from ι, are ο and ω, of the 4th magnitude, the two latter marking the shortest diagonal of a perfect diamond with β and θ — the latter being 10½° apart; 7° above ε is χ, of the 4th magnitude, and 6° above ε is δ, of the 3rd magnitude; these four form another diamond group of almost exactly the same size and shape as the former, and about 7° Northwest from it; χ is 14° East from α.

Sirius forms a line with γ, of the 2nd, δ, ε, and θ, the distances being : from Sirius to γ, 36°; γ to δ, 9°; δ to ε, 6°; and ε to θ, 11°; θ, ι, and μ, form a nearly equilateral triangle, averaging 18° on each side; ε and ι, in the South side of the Western diamond, form a trapezium with β and ο, farther South, almost identical in size and shape with the bowl of the Great Dipper; the Western diamond, with χ, 4° North from ε, forms a figure of five stars, similar to the group in the head of Draco, χ, at the acute angle, being in the Western end; North 16° from ι, is λ, of the 3rd magnitude, with a little

42   ASTRONOMY.

more than 9 hours of Right Ascension. The smaller stars in the constellation, including those in the sails, sometimes counted separately from Argo, may be easily located from the map.

129. Southeast by South 45° from Canopus, 20° a little North of East from β, East 11° from θ, and 12½° from τ, is α Crucis — of the 1st magnitude — the most Southerly star in the Southern Cross, situated just East of the hinder legs of the Centaur (Sec. 95); β Crucis, of the 3rd magnitude, and α, form a parallelogram with θ and χ Argûs, the longest side of which is parallel with the Equinoctial. East 30° from β Argûs, in line with Canopus, is Agena — β — and 5° East of β is α^2 — both of the 1st magnitude, and located in the front feet of the Centaur, whence α Centauri is sometimes called "Ungula" — the hoof.

South, 5° from α Crucis, are α and β Muscæ, of the 4th magnitude, forming with γ and δ, 5° farther South, a small trapezium which marks the location of the Southern Fly (Sec. 57). Southeast 9° from Ungula, and just outside the margin of the map, is γ Trianguli Australis, of the 3rd magnitude (Sec. 101). α Circini, of the 4th magnitude, between γ Trianguli and Ungula, is the only star of note in a small constellation known as "The Compasses." Below Canopus 11°, and 22° West from β Argûs, is β Doradûs (Sec. 73), of the 4th magnitude. East of South, 10° from Canopus, is α Pictoris, of the 4th magnitude, the only notable star in The Painter's Easel. West 24° from β Argûs, in line with Achernar, is γ Hydri, of the 3rd magnitude (Sec. 71); α Hydri is 16° farther in the same line; 13° West from γ Hydri, is β, of the 3rd magnitude; γ and β form an isosceles triangle with Achernar.

130. The region near the South Pole is singularly barren in stars visible with the naked eye; the star of the 5th magnitude, South of β Hydri, belongs to the constellation Octans — the Octant; it is the only star of greater magnitude than the 6th, near the Pole; and there are none exceeding it in apparent size, within a large area. The seemingly vacant space is filled up on the globe by several figures, as: Mons Mensæ (The Table Mountain), between Dorado and the Pole; Piscis Volans (The Flying Fish), just West of β Argûs; and the Chameleon, South of Muscæ and the stern of Argo. The nearest important stars, near the downward continuation of the perpendicular line on the map, are α, β, and γ Pavonis (Sec. 112), on the Antarctic circle; the middle star — β — has 20¼ hours of Right Ascension. The course of the Milky Way is very strongly marked in the region represented on this map; it runs from Canis Major, through the prow of the ship, and along the deck to the feet of the Centaur, which it includes, crossing several successive hour circles at nearly the same distance from the Pole; then turns Northward. The feet of the Centaur mark the Southern bend of the luminous band; it recedes from the South Pole on each side of Centaurus. From the Cross Westward, it runs through Argo, Monoceros, the feet of Gemini, the East part of Orion, Auriga, Perseus, Cassiopeia, and to the head of Cepheus. Here it divides, passing through Cygnus in two streams, which meet in the tail of Scorpio, and flow unitedly through Ara to Centaurus. One stream from Cygnus passes through Delphinus, Aquila, the feet of Antinous, Shield of Sobieski and the bow of Sagittarius to the tail of Scorpio; the other stream from Cygnus, through Anser (Sec. 132), Taurus Poniatowski, the side of Ophiuchus, and the tail of Scorpio.

DEFINE AND EXPLAIN (the figures refer to the sections):

127. Map No. XII; connections. 128. Argo, Tureis; Canopus; Naos; stern line; Robur Caroli; two diamond groups; Equilateral triangle. 129. Crux; Agena; Ungula; Muscæ; Triangulum Australis; Circinus; Doradus; Pictoris; Hydrus. 130. Octans, lack of stars; Course of the Milky Way.

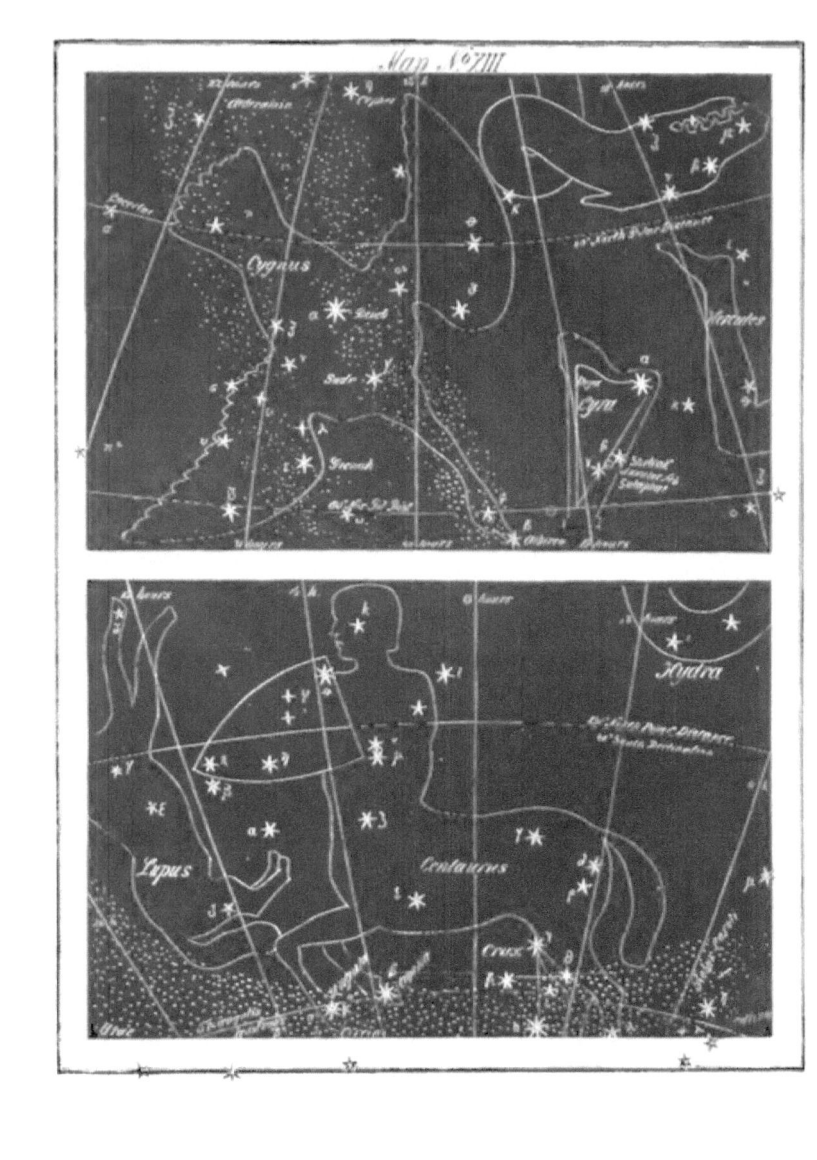

MAP NO. XIII.

131. This map is divided into two portions, the upper section showing a part of the Northern, the lower section a part of the Southern Hemisphere.

132. CYGNUS. The Swan. This constellation, represented in the upper part of the Map, is situated South of Cepheus, Southeast of the head of Draco, North of East from Lyra, and Northwest of Pegasus; it contains 81 discernible stars. Deneb — a — a brilliant white star of the 1st magnitude, with 20h 38m of Right Ascension, is 28° a little South of East from Etanin, 24° a little North of East from Vega, 18° a little West of South from Alderamin, 38° a little East of North from Altair, and 33° Northwest from Scheat Pegasi, in direct line with Algenib at the opposite corner of the Great Square. Its Declination is 44° 50, North Polar Distance 45° 10; it is hence nearly in the Zenith of the Northern States when on the Meridian.

Southwest 6° from a is Sadr — γ — a yellow colored star of the 3rd magnitude, and 16° farther, and the same distance from Vega, is Albireo — β — a topaz yellow in the beak, and of the 3rd magnitude; φ, of the 4th magnitude, is 2° from β towards a. γ, φ, and β, form an easily traced line in the heavens, which is crossed at γ by two stars of the 3rd magnitude, each 8° from γ — ϵ, Gienah, on the Southeast, and δ, on the Northwest, in line towards Etanin. Nearly in the same line, 7° Southeast from Gienah, is ζ, of the 3rd magnitude, in the Eastern wing. North 6° and a little West from δ, is θ, of the 4th magnitude, and 5° Northwest from θ is κ, of the 4th magnitude, near the first fold of Draco.

From Deneb, 6° towards θ, and forming a trapezium with a, γ, and δ, is o^2, of the 4th magnitude, Southeast 5° from a is ν, of the 4th magnitude; and 3° farther South, in the same line, and nearly East from Vega, is a star marked No. 61, in reality two stars of the 6th magnitude, revolving round each other, and forming the first discovered system of two sun-stars revolving about a common center. The stars marked a, τ, and ζ, near the North margin of the map, are in Cepheus.

Near the Eastern margin, with 40° of North Polar Distance, is a Lacertæ, of the 4th magnitude, the only star of note in the constellation called The Lizard; π^2 Pegasi, in the Southeast corner, is in the Northern hoof of Pegasus. Immediately South of Cygnus is a space between it and Aquila and Delphinus, which has been mapped out as belonging to Vulpecula et Anser (the Fox and Goose), and Sagitta (The Arrow). It contains no noteworthy stars, and the constellations are not, therefore, placed on these maps. Sagitta is an ancient constellation — the other a modern one.

The Milky Way runs through the section of the heavens represented in the upper half of this map. The Western stream from Ophiucus runs Northeast across the map, including the head and neck of Cygnus. The Eastern stream from Aquila takes the East wing of Cygnus in its course, from Gienah nearly to ζ, and the two streams run parallel across the body, uniting in the head of Cepheus.

44   ASTRONOMY.

133. CENTAURUS ET CRUX—The Centaur and the Cross. This constellation is represented in the lower section of Map No. XIII., the circle of 13 hours of Right Ascension being the central line. The Centaur is represented on the globe as a monster, half man and half horse, with shield and spear, attacking the Wolf in his front. This constellation contains 35 discernible stars, including 5 in the Cross, which is often spoken of as a separate constellation. The Centaur lies South of the tail of Hydra, Southwest of Libra and Scorpio, East of Argo, West of Ara, and East of the 12 hour circle; ι Centauri and β Lupi, both of the 3rd magnitude, and 4° apart, lying 24° Southwest from Antares (Sec. 99) are 13° East from μ and ν, both of the 3rd to the 4th magnitude, and are 4° East from ζ, of the 3rd magnitude; θ and ι, of the 3rd magnitude, 10° apart, in the shoulders, are on an East and West line, 6° and 7° above ν, and form with it an inverted triangle, the base of which is parallel with the Equinoctial; θ and ι form a diamond figure with η and ν, and the two latter form a lower quadrilateral with ζ, of the 3rd magnitude, 6° South from ν, and α Lupi 6° below η. These, with k, of the 4th magnitude, in the head, and a few smaller stars in the shield, are all of the Centaur, visible North of the Gulf of Mexico. The most prominent stars in the constellation never rise above the horizon of the United States.

East 11° from θ Argûs, is α Crucis, of the 1st magnitude, and 6° North from α, is γ, of the 2nd magnitude, these two forming the upright portion of the Cross, situated a little East of the 12 hour circle, and pointing to the South Pole, which is 27½° from α; 4° apart, and nearly perpendicular to these, are β, of the 2nd magnitude, and δ, of the 3rd, the latter being nearly on the 12 hour circle,

and with δ Centauri, of the 3rd magnitude, 8° Northward, lying on the Equinoctial Colure, with α Corvi, δ Ursæ Majoris, β Cassiopeiæ, α Andromedæ, γ Pegasi, and β Hydri, the latter being on the other side of the South Pole from the Cross; δ is 18° West from ζ; and 6° East from δ towards ζ, is γ, of the 3rd magnitude; ϵ, of the 3rd magnitude, is nearly half way between ζ and β Crucis, α Crucis is 11° East from θ Argûs, and 12° farther East is Agena — β Centauri — of the 1st magnitude, in the second fore foot; 5° farther East, and 33° from β Argûs, 9° West from β Trianguli, and 22° West from β Aræ (Sec. 100), is α^2 Centauri, of the 1st magnitude — Ungula — close to this is α^1, of the 4th magnitude, both in the forward hoof of Centaurus.

134. LUPUS—The Wolf. This constellation lies East of Centaurus, the head towards the North, and nearly touching the body of Scorpio; it contains 24 discernible stars. Southwest 6° from β, of the 2nd magnitude (Sec. 99), just South of ι Centauri, is α, of the 3rd magnitude, forming an isosceles triangle with β Lupi and η Centauri. Southeast 7° from α is ζ, of the 4th magnitude; East from ι Centauri, and 12° above ζ, is γ, of the 3rd magnitude. 13° North of ι Centauri, towards Zubenelg is δ, of the 4th magnitude, in the fore paw; Northeast from γ, and 2½° apart, are θ and η, both of the 4th magnitude, in the face. These are all the stars worthy of note in the constellation.

The Milky Way skirts the Southern limit of the lower section of the map, its Northern edge passing through the middle of Crux, and just outside of the fore feet of the Centaur. It connects with Argo West of the Centaur, and with Ara to the East.

DEFINE AND EXPLAIN (the figures refer to the sections):
131. Map No. XIII. 132. Cygnus; location; Deneb; Sadr; Albireo; Gienah; the Cross; remarkable star; the Lizard; Vulpecula et Anser; Sagitta; Milky Way. 133. Centaurus et Crux; Lupus; Shoulder line; shield; The Southern Cross; Equinoctial Colure; Ungula. 134. Lupus.

TABLE OF FIXED STARS.

135. The following table contains the places of the principal fixed stars, including all of the 1st, 2nd, and 3rd magnitudes, those of the 4th within the Zodiacal constellations, and in other groups in which there is no more prominent star; also a few of the 5th magnitude, and the Beehive nebula. The places are given in Right Ascension, North Polar Distance (Sec. 20), Longitude, and Latitude, for the beginning of the year 1875. The annual variation in Right Ascension and North Polar Distance, is also given for each star; marked + when the correction should be added for future years, and — when the correction should be subtracted from the tabular quantity for each subsequent year. For years previous to 1875, subtract where marked +, and add where the minus sign is given. Astronomers prefer the notation in North Polar Distance to that of Declination, in the tables, as it preserves a greater degree of uniformity in the signs (plus and minus) of the variation. The annual variation in Longitude of all the stars, is + 50¼′. The Latitudes do not vary.

The Longitudes and Latitudes are given to the nearest 1/10th of a minute of Arc.

MAG.	STAR.		RT. ASCEN.	ANN. VAR.	N. P. D.	ANN. VAR.		LONG.	LAT.
1	α Androm.;	*Alpheratz.*	0 1m 16s	+ 3.01s	61° 35′ 58″	— 19.9″	♈	12° 34.1′	N 25° 41.0′
2-3	β Cassiopeiæ;	*Chaph.*	2m 31s	3.16	31° 32′ 23″	19.9	♉	3° 21.2′	N 51° 12.6′
2	γ Pegasi;	*Algenib.*	6m 48s	3.08	75° 30′ 41″	20.0	♈	7° 24.9′	N 12° 36.8′
3	β Hydri.		19m 09s	3.27	167° 57′ 38″	20.3	♑	20° 5.3′	S 64° 41.1′
2	α Phœnicis.		20m 06s	2.98	132° 58′ 59″	19.7	♓	13° 44.1′	S 40° 36.7′
3	δ Andromedæ.		32m 39s	3.19	59° 49′ 25″	19.8	♈	20° 4.3′	N 24° 20.6′
3	α Cassiopeiæ;	*Schedir.*	33m 26s	3.37	34° 8′ 54″	19.8	♉	6° 3.1′	N 46° 36.8′
2-3	β Ceti;	*Diphda.*	37m 19s	3.02	108° 40′ 23″	19.9	♈	0° 49.3′	S 20° 47.0′
5	δ Piscium.		42m 12s	3.11	83° 5′ 43″	19.7	♈	12° 24.0′	N 2° 10.4′
5	γ Cassiopeiæ.		49m 11s	3.56	29° 57′ 37″	19.6	♉	12° 12.0′	N 48° 48.1′
4	ε Piscium.		56m 28s	3.11	82° 47′ 0″	19.5	♈	15° 47.0′	N 1° 5.1′
3-4	β Phœnicis.	I	0m 30s	2.69	137° 23′ 17″	19.4	♓	18° 40.6′	S 48° 12.0′
2	β Andromedæ;	*Mirach.*	2m 44s	3.34	55° 2′ 31″	19.3	♈	28° 39.8′	N 25° 56.1′
2	α Urs. Min.;	*Alruccabah.*	12m 55s	20.40	1° 21′ 25″	19.1	♊	26° 49.2′	N 66° 5.3′
3	δ Cassiopeiæ;	*Ruchbah.*	17m 39s	3.87	30° 21′ 54″	18.9	♉	16° 41.0′	N 46° 23.8′
3	θ Ceti.		17m 40s	3 00	98° 49′ 43″	18.7	♈	14° 29.0′	S 15° 46.0′
3	γ Phœnicis.		22m 57s	2.63	133° 57′ 29″	18.6	♓	26° 23.3′	S 47° 34.8′
4-5	η Piscium.		23m 38s	3.14	81° 30′ 8″	18.6	♈	21° 22.7′	S 3° 4.0′
4	ζ Piscium.		24m 48s	3.20	75° 17′ 56″	18.8	♈	25° 4.3′	N 5° 22.1′
3-4	51 Andromedæ.		1 30m 20s	+ 3.65	42° 0′ 22″	— 18.4	♉	10° 42.2′	N 35° 21.3′

ASTRONOMY.

MAG.	STAR.	RT. ASCEN.	ANN. VAR.	N. P. D.	ANN. VAR.	LONG.	LAT.
1	α Eridani; *Achernar.*	I 33m 3s +	2.24s	147° 52' 19" —	18.4s	♓ 13° 31.9'	S 59° 22.3'
3	ζ Ceti; *Baten Kaitos.*	45m 17s	2.96	100° 57' 1,"	17.0	♈ 20° 11.9'	S 20° 20.5'
3	ε Cassiopeiæ.	45m 25s	4.24	26° 56' 47"	18.0	♉ 23° 1.7'	N 47° 32.0'
3-4	α Trianguli.	45m 58s	3.40	61° 1' 50"	17.8	5° 7.4'	N 16° 47.9'
3	β Arietis; *Sheratan.*	47m 44s	3.30	69° 48' 13"	17.8	♉ 2° 13.5'	N 8° 28.9'
3	α Hydri.	54m 50s	1.89	152° 10' 44"	17.6	♓ 10° 19.1'	S 61° 13.8'
3-4	α Piscium; *El Rischa.*	55m 35s	3.10	87° 50' 21"	17.6	♈ 27° 37.8'	S 9° 4.2'
3	γ Andromedæ; *Alamach.*	56m 14s	3.65	48° 10' 15"	17.6	♉ 12° 29.1'	N 27° 47.7'
2	α Arietis; *Hamal.*	II 0m 8s	3.36	67° 7' 48"	17.3	5° 54.9'	N 9° 57.7'
Var.	ο Ceti; *Mira.*	13m 2s	3.02	93° 32' 47"	16.6	♈ 29° 46.6'	S 15° 56.7'
4	35 Arietis.	36m 7s	3.50	62° 45' 30"	15.7	♉ 15° 11.5'	N 11° 18.6'
3	γ Ceti.	36m 49s	3.10	87° 17' 32"	15.4	7° 41.6'	S 12° 0.3'
4	μ Arietis.	38m 11s	3.23	80° 21' 53"	15.5	10° 19.0'	S 4° 17.6'
5	π Arietis.	42m 19s	3.33	79° 3' 22"	15.3	13° 22.4'	N 1° 48.7'
3	41 Arietis.	42m 38s	3.51	63° 15' 21"	15.2	16° 27.6'	N 10° 26.3'
3	τ Eridani; *Azha.*	50m 19s	2.93	99° 23' 47"	14.6	6° 59.9'	S 24° 33.0'
2-3	α Ceti; *Menkar.*	55m 45s	3.13	86° 24' 7"	14.4	12° 31.4'	S 12° 35.8'
3-4	γ Persei.	55m 45s	4.31	36° 59' 6"	14.4	28 16.9	N 31° 30.8'
2-3	β Persei; *Algol.*	III 0m 2s	3.87	49° 31' 36"	14.3	24° 25.6'	N 22° 24.8'
4	δ Arietis.	4m 29s	3.42	70° 44' 50"	14.0	♉ 19° 6.1'	N 1° 48.6'
2-3	α Persei; *Mirfak.*	15m 24s	4.24	40° 35' 9"	13.1	♊ 0° 20.3'	N 30° 6.6'
4-5	10 Tauri.	30m 30s	3.06	89° 59' 47"	11.7	♉ 20° 13.2'	S 18° 26.1'
3	δ Persei.	34m 2s	4.24	42° 36' 51"	11.9	♊ 3° 3.5'	N 27° 17.1'
3-4	δ Eridani.	37m 16s	2.87	100° 11' 15"	12.5	♉ 19° 6.4'	S 28° 43.4'
5	19 Tauri.	37m 46s	3.56	65° 55' 34"	11.7	27° 49.2'	N 4° 30.2'
3	η Tauri; *Alcyone.*	40m 3s	3.55	66° 15' 58"	11.6	♉ 28° 11.9'	N 4° 2.2'
3-4	ξ Persei.	46m 17s +	3.75	58° 29' 22"	11.0	♊ 1° 22.8'	N 11° 19.0'
3	γ Hydri.	49m 14s —	1.01	164° 37' 21"	10.9	≏ 4° 41.0'	S 76° 21.3'
2-3	γ Eridani; *Zaurak.*	52m 12s +	2.80	103° 51' 35"	10.6	♉ 22 7.6'	S 33° 12.8'
4	λ Tauri.	53m 45s	3.31	77° 51' 51"	10.6	♉ 28° 53.3'	S 7° 58.5'
3-4	γ Tauri.	IV 12m 41s	3.41	74° 40' 34"	9.0	♊ 4° 4.4'	S 5° 44.7'
3-4	α Reticuli.	12m 19s	0.74	152° 47' 18"	9.0	♈ 5° 39.6'	S 78° 2.8'
4	θ¹ Tauri.	15m 44s	3.45	72° 45' 9"	8.8	♊ 5° 7.3'	S 3° 50.0'
3-4	ε Tauri.	21m 19s	3.49	71° 5' 55"	8.4	♊ 6° 42.9'	S 2° 34.9'
1	α Tauri; *Aldebaran.*	28m 43s	3.44	73° 44' 37"	7.6	♊ 8° 1.5'	S 5° 28.6'
3-4	ο¹ Eridani; *Theemin.*	30m 42s	2.33	120° 49' 12"	7.7	♉ 28° 8.1'	S 51° 50.1'
3	α Doradûs.	31m 18s	1.29	145° 18' 15"	7.6	♉ 6° 2.0'	S 74° 35.6'
4	α Camelopardi.	41m 12s	5.92	23° 54' 22"	6.7	♊ 19° 10.1'	N 43° 22.6'
4-5	ι Tauri.	55m 37s	3.58	68° 35' 26"	5.5	15° 2.4'	S 1° 12.9'
3	β Eridani; *Cursa.*	V 1m 12s	2.95	95° 14' 58"	5.1	13° 36.0'	S 27° 53.6'
5	15 Orionis.	2m 33s	3.43	74° 33' 50"	5.0	16° 3.0'	S 7° 19.7'
1	α Aurigæ; *Capella.*	7m 27s	4.42	44° 7' 54"	4.1	20° 6.8'	N 22° 51.8'
1	β Orionis; *Rigel.*	8m 32s	2.88	98° 20' 51"	4.6	15° 5.0'	S 31° 8.1'
2	β Tauri; *El Nath.*	18m 24s	3.79	61° 36' 2"	3.4	20° 50.0'	N 5° 22.4'
2	γ Orionis; *Bellatrix.*	5 18m 26s +	3.22	83° 45' 56" —	3.6	♊ 19 12.1'	S 16 50.0'

TABLE OF FIXED STARS. 47

MAG.	STAR.		RT. ASCEN.	ANN. VAR.	N. P. D.	ANN. VAR.		LONG.	LAT.
2	δ Orionis; *Mintaka*.	V	25m 37s +	3.07s	90° 23′ 38″ —	3.0s	Π	20° 37.0′	S 23° 31.3′
3-4	α Leporis; *Arneb*.		27m 13s	2.65	107° 54′ 47″	3.0		19° 38.1′	S 41° 4 .6′
2-3	ε Orionis; *Alnilam*.		29m 52s	3.04	91° 17′ 1″	2.6		21° 43.1′	S 24° 31.5′
3-4	ζ Tauri.		30m 11s	3.58	68° 56′ 9″	2.6		22° 2.4′	S 1° 45.5′
3	ζ Orionis; *Alnitak*.		34m 27s	3.03	92° 0′ 40″	2.2		22° 56.2′	S 25° 18.7′
2	α Columbæ; *Phact*.		35m 8s	2.18	124° 8′ 32″	2.2		20° 25.′5	S 57° 23.6′
3	κ Orionis; *Saiph*.		41m 50s	2.85	99° 42′ 57″	1.7		24° 39.2′	S 33° 5.2′
4-5	136 Tauri.		45m 28s	3.77	62° 25′ 13″	1.2		26° 46.4′	N 4° 9.8′
3	β Columbæ.		46m 33s	2.11	125° 49′ 5″	1.5		24° 40.3′	S 59° 12.9′
1	α Orionis; *Betelguese*.		48m 24s	3.25	82° 37′ 6″	1.0		27° 0.5′	S 16° 2.7′
3-4	δ Aurigæ.		49m 14s	4.94	35° 43′ 40″	0.8		28° 10.2′	N 30° 50.2′
2	β Aurigæ; *Menkalinan*.		50m 22s	4.40	45° 4′ 5″ —	0.8	Π	28° 10.0′	N 21° 29.4′
4	η Geminorum; *Tejat*.	VI	7m 20s	3.62	67° 27′ 33″ +	0.6	♋	1° 41.7′	S 0° 54.3′
3	μ Geminorum.		15m 24s	3.64	67° 25′ 28″	1.5		3° 33.4′	S 0° 50.0′
2-3	ξ Canis Majoris; *Phurid*.		15m 31s	2.30	120° 0′ 34″	1.3		5° 38.3′	S 53° 23.5′
2-3	β Canis Majoris; *Mirzam*.		17m 12s	2.64	107° 53′ 43″	1.4		5° 26.8′	S 41° 16.3′
1	α Argûs; *Canopus*		21m 11s	1.33	142° 37′ 41″	1.8		13° 14.4′	S 75° 50.3′
4	ν Geminorum.		21m 32s	3.57	69° 42′ 39″	1.9		5° 1.8′	S 2° 17.5′
2-3	γ Geminorum; *Alhena*.		30m 29s	3.47	73° 29′ 45″	2.7		7° 21.5′	S 6° 45.6′
3	ν Argûs.		33m 56s	1.83	133° 5′ 10″	2.8		15° 21.3′	S 66° 5.8′
3	ε Geminorum; *Mebsuta*.		36m 15s	3.70	64° 44′ 51″	3.2		8° 11.6′	N 2° 7.4′
4	ξ Geminorum.		38m 17s	3.37	76° 58′ 18″	3.5		9° 28.2′	S 10° 6.8′
1	α Canis Majoris; *Sirius*.		39m 38s	2.65	106° 32′ 43″	4.5	♋	12° 21.0′	S 39° 34.2′
4	α Pictoris.		46m 54s	0.61	151° 48′ 28″	3.8	♌	22° 33.7′	S 83° 3.4′
2-3	ε Canis Majoris; *Adhara*.		53m 43s	2.36	118° 48′ 11″	4.6	♋	19 2.7′	S 51° 22.6′
4	ζ Geminorum.		56m 42s	3.57	69° 14′ 54″	4.9		13° 14.8′	S 2° 3.4′
3-4	δ Canis Maj.; *Wesen*.	VII.	3m 19s	2.44	116° 11′ 45″	5.4		21° 39.7′	S 48° 28.2′
4-5	λ Geminorum.		10m 55s	3.46	73° 14′ 9″	6.1		17° 2.2′	S 5° 39.0′
3	δ Geminorum; *Wasat*.		12m 40s	3.60	67° 47′ 23″	6.3		16° 46.5′	S 0° 11.8′
3	π Argûs.		12m 45s	2.14	126° 52′ 26″	6.2		28° 35.2′	S 58° 32.8′
4	ι Geminorum.		17m 58s	3.74	61° 57′ 20″	6.8		17° 13.0′	N 5° 44.8′
2	η Canis Majoris; *Alwlra*.		19m 9s	2.37	119° 3′ 39″	6.7		27° 48.3′	S 50° 37.6′
3	β Canis Min.; *Gomeisa*.		20m 22s	3.26	81° 27′ 37″	6.9		20° 27.1′	S 15° 30.1′
1-2	α²Geminorum; *Castor*.		26m 37s	3.85	57° 50′ 22″	7.5		18° 30.1′	N 10° 5.1′
1	α Canis Min.; *Procyon*.		32m 45s	3.14	84° 27′ 19″	8.8		24° 3.7′	S 15° 58.1′
4	κ Geminorum.		36m 54s	3.03	65° 18′ 15″	8.3		21° 55.3′	N 3° 1.0′
2	β Geminorum; *Pollux*.		37m 46s	3.68	61° 40′ 26″	8.3	♋	21° 29.4′	N 6° 40.5′
2-3	ζ Argûs; *Naos*.		39m 12s	2.11	129° 39′ 10″	10.0	♌	16° 50.1′	S 58° 22.5′
3-4	ρ Argûs; *Tureis*.	VIII.	2m 13s	2.56	113° 56′ 41″	10.0	♌	9° 39.7′	S 43° 17.1′
4	φ²Cancri.		2m 55s	3.03	64° 6′ 52″	10.6	♌	27° 30.3′	N 0° 32.0′
2	γ Argûs.		5m 40s	1.84	136° 58′ 9″	10.5	♌	25° 38.4′	S 64° 28.3′
4	β Cancri.		9m 44s	3.26	80° 25′ 50″	10.7	♌	2° 31.6′	S 10° 17.9′
2	ε Argûs.		10m 57s	1.24	149° 6′ 24″	11.3	♍	21° 23.7′	S 72° 36.2′
4	δ Hydræ.		31m 2s	3.18	83° 51′ 40″	12.2	♌	8° 33.9′	S 12° 24.2′
Neb Cancri; *Præsepe*.		8	33m 16s +	3.48	69° 54′ 0″ +	12.5	♌	5° 33.6′	N 1° 14.0′

48   ASTRONOMY.

MAG.	STAR.		RT. ASCEN.	ANN. VAR.	N. P. D.	ANN. VAR.		LONG.		LAT.
4–5	γ Cancri; *N. Asell.*	VIII	36m 3s	+ 3.49s	68° 4' 57"	+ 12.5s	♌	5° 47.8'	N	3° 10.7'
4–5	δ Cancri; *Sou. Asellus.*		37m 35s	3.42	71° 23' 1.'	12.9	♌	6° 58.5'	N	0° 4.5'
3	δ Argûs.		41m 15s	1.65	141° 15' 5"	13.1	♍	17° 14.3'	S	67° 11' 8"
3–4	ε Ursa Majoris; *Talita.*		50m 3'8	4.11	41° 28' 10"	13.9	♌	1° 3.9'	N	29° 34.5'
4	α Cancri; *Acubens.*		51m 39s	3.29	77° 39' 31"	13.6	♌	11° 53.8'	S	5° 5.5'
3	λ Argûs.	IX	3m 23s	2.26	132° 55' 46"	14.4	♍	9° 27.9'	S	55° 53.3'
4–5	θ Hydrae.		7m 52s	3.13	87° 9' 32"	14 9	♌	18° 32.3'	S	13° 3.1'
1	β Argûs.		11m 49s	0.68	159° 12' 11'	14.8	♍	0° 17.4'	S	72° 13.1'
4	α Lyncis; (*The Lynx*)		13m 26s	3.68	55° 4' 30"	14.9	♌	10° 6.6'	N	17° 57.2'
2	ι Argûs.		14m 15s	1.60	148° 45' 2"	14.9	♎	3° 37.6'	S	67° 4.7'
3	ι Argûs.		18m 15s	1.85	144° 28' 42"	15.3	♍	27° 9.1'	S	63° 57.5'
2	α Hydrae; *Alphard.*		21m 27s	2.95	98° 7' 4"	15.3	♌	25° 32.5'	S	22° 23.5'
3	θ Ursae Majoris.		24m 29s	4.63	37° 15' 16"	16.1		5° 36.4'	N	34° 53.8'
4	ω Leonis.		34m 29s	3.23	79° 32' 24"	16.1		22° 30.5'	S	5° 45.8'
3	ε Leonis.		38m 45s	3.42	65° 39' 4"	16.3	♌	17° 55.2'	N	9° 24.0'
3	υ Argûs		43m 59s	1.50	154° 29' 33"	16.6	♎	21° 11.1'	S	67° 29.6'
3	μ Leonis; *Rasal Asad.*		45m 39s	3.63	64° 24' 10"	16.7	♌	19° 41.4'	N	12° 20.8'
1	η Leonis.	X	0m 31s	3.28	72° 37' 43"	17.3		26° 9.5'	N	4° 51.4'
1	α Leonis (*Cor*); *Regulus.*		1m 43s	3.21	77° 25' 22"	17.4		28° 5.2'	N	0° 27.6'
4–5	ζ Leonis; *Althafara.*		9m 14s	3.35	65° 57' 36"	17.7		25° 19.0'	N	11° 51.3'
2	γ Leonis; *Algeiba.*		13m 5s	3.32	69° 31' 37"	18.0		27° 51.2'	N	8° 48.0'
3	μ Urs. Maj.; *El Pheckra.*		14m 53s	3.60	47° 52' 21"	17.9		19° 29.1'	N	28° 59.3'
4–5	β Leonis Minoris.		20m 39s	3.50	52° 39' 10"	18.3	♌	22° 17.4'	N	25° 3.4'
3	θ Argûs.		38m 30s	2.12	153° 41' 23"	18.8	♎	27° 28.5'	S	62° 7.8'
2	η Argûs.		40m 13s	2.30	149° 1' 38"	18.7		26° 25.8'	S	58° 54.9'
3	μ Argûs.		44m 24s	2.56	138° 45' 34"	19.0	♎	8° 17.1'	S	51° 5.2'
4	α Crateris; *Alkes.*		53m 31s	2.92	107° 37' 58"	19.0	♍	21° 1.8'	S	23° 6.4'
2	Urs. Maj.; *Merak.*		54m 17s	3.67	32° 56' 53"	19 2	♌	17° 40.4'	N	45° 7.1'
1–2	α Urs. Maj.; *Dubhe.*		56m 0s	3.76	27° 31' 29"	19.3	♌	13° 20.4'	N	49° 36.5'
4	β Crateris.	XI	5m 31s	2.95	112° 8' 36"	19.6	♍	26° 19.0'	S	25° 38.4'
2–3	δ Leonis; *Zozma.*		7m 28s	3.21	68° 47' 30"	19.6		9° 32.8'	N	14° 20.2'
3	θ Leonis.		7m 41s	3.16	73° 53' 9"	19.5		11° 40.5'	N	9° 40.6'
3–4	δ Crateris.		13m 6s	3.00	104° 6' 9"	19.4		24° 57.4'	S	17° 34.7'
4	σ Leonis.		14m 41s	3.10	83° 17' 10"	19.7		16° 57.8'	N	1° 41.8'
4	ι Leonis.		17m 24s	3.14	78° 46' 35"	19.7		15° 18.6'	N	6° 6.2'
4	τ Leonis.		21m 31s	3.09	86° 27' 20"	19.8		19° 46.0'	S	0° 33.3'
4–5	ε Leonis.		23m 56s	3.07	92° 18' 52"	19.8	♍	22° 37.6'	S	5° 41.6'
3–4	λ Draconis; *Giansar.*		23m 5s	3.64	19° 58' 47"	19.9	♌	8° 34.2'	N	57° 13.8'
4–5	ν Virginis.		39m 26s	3.09	82° 46' 13"	20.2	♍	22° 21.8'	N	4° 36.1'
2–3	β Leonis; *Denebola.*		42m 41s	3.07	74° 13' 45"	20.1		19° 53.2'	N	12° 17.2'
3–4	β Virginis; *Zavijava.*		44m 11s	3.13	87° 31' 52"	20.3	♍	25° 23.2'	N	0° 41.6'
2	γ Urs. Maj.; *Phecda.*		47m 15s	3.19	35° 36' 38"	20 0	♌	28° 12.5'	N	47° 8.0'
3	δ Centauri.	XII	1m 53s	3.08	110° 1' 38"	20.1	♎	25° 15.3'	S	41° 29.9'
4–5	α Corvi; *Alchiba.*		1m 58s	3.08	114° 1' 51"	20.1	♎	10° 30.1'	S	21° 41.8'
3	δ Crucis.	12	7m 31s	+ 3.14	148° 3' 9"	+ 20.0	♏	3° 56.1'	S	50° 24.5'

TABLE OF FIXED STARS.

MAG.	STAR.		RT. ASCEN.	ANN. VAR.	N. P. D.	ANN. VAR.		LONG.		LAT.	
3	δ Urs. Maj.; *Megrez.*	XII	9m 14s	+ 3.018	32° 16' 24"	+ 20.1s	♌	29° 17.9'	N	51° 38.9'
3	γ Corvi.		9m 23s	3.08	106° 50' 51"	20.0	♎	8° 59.7'	S	14° 29.9'
3-4	η Virginis.		13m 31s	3.07	89° 58' 20"	20.1	♎	3° 5.3'	N	1° 22.2'
1	α Crucis.		19m 39s	3.27	152° 24' 18"	19.9	♏	10° 8.5'	S	52° 51.7'
3	δ Corvi; *Algorab.*		23m 24s	3.11	105° 19' 10"	20.1	♎	11° 43.1'	S	12° 10.9'
2	γ Crucis.		24m 15s	3.29	146° 24' 41"	20.1	♏	4° 59.9'	S	47° 48.6'
2-3	β Corvi.		27m 49s	3.13	112° 42' 20"	20.0	♎	15° 37.6'	S	18° 2.1'
4	α Muscæ.		29m 45s	3.50	158° 26' 46"	19.9	♏	18° 38.7'	S	56° 32.4'
3	γ Centauri.		34m 38s	3.28	138° 16' 24"	19.9	♏	0° 35.6'	S	40° 8.8'
4	γ Virginis.		35m 20s	3.04	90° 45' 50"	19.8	♎	8° 25.0'	N	2° 48.2'
2	β Crucis.		40m 26s	3.46	149° 0' 13"	19.7	♏	9° 55.0'	S	48° 37.4'
3	ε Urs. Maj.; *Alioth.*		48m 32s	2.67	33° 21' 43"	19.7	♍	7° 9.8'	N	54° 18.5'
3	δ Virginis.		49m 18s	3.02	85° 55' 24"	19.7	♎	9° 43.1'	N	8° 37.0'
2-3	α Can. Ven.; *Cor Caroli.*		50m 11s	2.82	51° 0' 23"	19.6	♍	22° 49.0'	N	40° 7.5'
3	ε Virginis; *Vindemiatrix.*		55m 58s	2.99	78° 22' 5"	19.5	♎	8° 12.1'	N	16° 12.9'
4-5	α Comæ Berenices.	XIII	3m 58s	2.92	71° 48' 39"	19.2		7° 13.1'	N	22° 59.1'
4	γ Hydræ.		12m 8s	3.25	112° 30' 37"	19.1	♎	25° 16.4'	S	13° 43.9'
3	ι Centauri.		13m 35s	3.35	126° 3' 14"	19.2	♏	1° 21.2'	S	25° 59.7'
1	α Virginis (Spica); *Arista.*		18m 37s	3.15	100° 30' 31"	19.0	♎	22° 5.9'	S	2° 2.6'
3	ζ Ursæ Majoris; *Mizar.*		18m 54s	2.44	34° 25' 18"	18.9	♍	13° 55.7'	N	50° 21.6'
4	ζ Virginis.		28m 20s	3.05	89° 57' 22"	18.6	♎	20° 8.8'	N	9° 16.0'
3	ε Centauri.		31m 59s	3.75	142° 49' 18"	18.6	♏	13° 49.4'	S	39° 34.3'
3-4	ν Centauri.		42m 2s	3.57	131° 3' 52"	18.2	♏	9° 25.0'	S	28° 15.2'
2-3	η Urs. Majoris; *Alkaid.*		42m 36s	2.35	40° 3' 45"	18.1	♍	25° 10.1'	N	54° 23.5'
3	ζ Centauri.		47m 45s	3.70	136° 40' 21"	18.0	♏	13° 12.9'	S	32° 55.6'
3	η Bootis; *Muphrid.*		48m 44s	2.86	70° 58' 30"	18.2	♎	17° 31.8'	N	28° 6.1'
1	β Centauri; *Agena.*		55m 1s	4.17	149° 16' 10"	17.7	♏	22° 3.5'	S	44° 7.2'
2-3	θ Centauri.		59m 19s	3.50	125° 45' 22"	18.1	♏	10° 31.2'	S	22° 2.2'
3-4	α Draconis; *Thuban.*	XIV	1m 0s	1.62	25° 1' 56"	17.4	♍	5° 40.5'	N	66° 21.5'
4	ι Virginis.		6m 14s	3.19	99° 41' 32"	17.1	♏	2° 15.0'	N	2° 55.1'
4	ι Virginis.		9m 28s	3.14	95° 24' 10"	17.4	♏	2° 2.7'	N	7° 13.8'
1	α Bootis; *Arcturus.*		9m 58s	2.73	70° 9' 57"	18.9	♎	22° 29.3'	N	30° 49.5'
4	λ Virginis.		12m 21s	3.24	102° 17' 42"	16.8	♏	5° 12.5'	N	0° 30.1'
3-4	γ Bootis; *Seginus.*		27m 3s	2.43	51° 8' 40"	16.0	♎	15° 54.6'	N	49° 33.3'
3	η Centauri.		27m 35s	3.78	131° 36' 31"	16.2	♏	18° 30.6'	S	25° 29.8'
1	α² Centauri; *Ungula.*		31m 8s	4.04	150° 18' 55"	15.1	♏	27° 56.6'	S	42° 32.3'
4	α Circini.		32m 27s	4.77	154° 25' 45"	16.1	♐	0° 37.9'	S	46° 10.5'
3	α Lupi.		33m 38s	3.95	136° 51' 3"	15.9	♏	21° 5.9'	S	30° 1.2'
3-4	ζ Bootis.		35m 8s	2.86	75° 44' 4"	15.7		1° 17.0'	N	27° 53.6'
4-5	μ Virginis.		36m 28s	3.15	95° 6' 51"	16.0	♏	8° 22.5'	N	9° 11.7'
3	ε Bootis; *Izar.*		39m 32s	2.62	62° 23' 54"	15.4	♎	26° 21.0'	N	40° 38.2'
3	α² Libræ; *Zubenesch.*		43m 58s	3.31	105° 31' 17"	15.3	♏	13° 20.9'	N	0° 20.0'
3	β Lupi.		50m 21s	3.90	132° 37' 17"	15.0		23° 17.2'	S	25° 1.7'
3	κ Centauri.		51m 2s	+ 3.87	131° 36' 3"	14.8	♏	23° 3.3'	S	24° 0.8'
3	β Ursæ Minoris; *Kochab.*		51m 5¼s	− 0.25	15° 20' 3"	+ 14.8	♌	11° 32.4'	N	72° 58.5'

50   ASTRONOMY.

MAG.	STAR.		RT. ASCEN.	ANN. VAR.	N. P. D.	ANN. VAR.		LONG.		LAT.
4–5	δ Libræ.	XIV	54m 18s	+ 3.20s	98° 1′ 19″	+ 14.68	♏	13° 32.3′	N	8° 15.6′
4–4	20 Libræ.		56m 45s	3.50	111° 47′ 22″	14.5	♏	18° 56.6′	S	7° 37.3′
3	β Boötis; Nekkar.		57m 14s	2.26	49° 6′ 57″	14.5	♎	22° 29.1′	N 54° 9.7′
3	γ Triang. Australis.	XV.	7m 16s	5.49	158° 12′ 56″	13.8	♐	7° 39.4′	S 48° 5.0′
2–3	β Libræ; Zubenelg.		10m 17s	3.22	98° 55′ 14″	13.6	♏	17° 37.7′	N 8° 30.8′
3–4	δ Boötis; Alkaturgos.		10m 28s	+ 2.42	56° 13′ 4″	13.7	♏	1° 23.3′	N 48° 58.8′
3–4	γ Ursæ Minoris.		20m 57s	− 0.13	17° 43′ 15″	12.8	♌	19° 47.9′	N 75° 13.9′
4	ζ² Libræ.		21m 13s	+ 3.37	106° 16′ 46″	12.0	♏	22° 9.2′	N 2° 6.4′
3	ε Draconis.		22m 9s	1.33	30° 35′ 44″	12.8	♎	3° 9.3′	N 71° 5.8′
3	γ Lupi.		26m 49s	3.96	130° 44′ 46″	12.6	♏	29° 45.4′	S 21° 13.6′
4	37 Libræ.		27m 21s	3.27	99° 38′ 5″	12.7		21° 55.6′	N 8° 55.3′
4–5	γ Libræ.		28m 32s	3.35	104° 22′ 17″	12.4		23° 23.4′	N 4° 21.0′
3	δ Serpentis.		28m 50s	2.87	79° 2′ 28″	12.3		16° 35.7′	N 28° 53.8′
2–3	α Cor. Borealis; Alphecca.		29m 24s	2.54	62° 51′ 49″	12.4		10° 31.9′	N 44° 20.3′
1	39 Libræ.		29m 26s	3.62	117° 43′ 10″	12.3		26° 52.0′	S 8° 29.4′
4–5	η Libræ.		37m 3s	3.37	105° 16′ 24″	11.9		25° 36.5′	N 4° 1.1′
2–3	α Serpentis; Unukalhay.		38m 7s	2.95	83° 10′ 48″	11.7		20° 10.0′	N 25° 31.0′
3–4	β Serpentis.		40m 25s	2.77	74° 11′ 5″	11.6	♏	18° 11.5′	N 31° 20.6′
3	β Triangulis Australis.		44m 9s	5.23	153° 2′ 28″	11.6	♐	10° 6.3′	S 41° 54.6′
3	ε Serpentis.		44m 35s	2.99	85° 8′ 41″	11.2	♏	22° 34.4′	N 24° 1.9′
4	λ Libræ.		46m 5s	3.47	109° 47′ 30″	11.2	♏	28° 43.9′	N 0° 6.2′
4	ρ Scorpii.		49m 10s	3.69	118° 50′ 51″	10.9	♐	1° 15.3′	S 8° 36.2′
3	γ Serpentis.		50m 41s	2.77	73° 55′ 12″	12.0	♐	20° 59.9′	N 35° 15.6′
3–4	π Scorpii.		51m 17s	3.61	115° 45′ 9″	10.7	♐	1° 11.7′	S 5° 27.4′
3	δ Scorpii.		52m 57s	3.53	112° 15′ 50″	10.6	♐	0° 49.6′	S 1° 58.1′
4–5	51 Libræ.		57m 30s	3.29	101° 1′ 37″	10.3	♏	29° 33.7′	N 9° 15.2′
2	β²Scorpii; Graffias.		58m 10s	3.18	109° 27′ 41″	10.2	♐	1° 26.7′	N 1° 1.4′
3	θ Draconis.		59m 33s	1.12	31° 6′ 3″	9.8	♎	14 56.0′	N 74° 26.0′
4	ν Scorpii.	XVI	4m 44s	3.48	109° 8′ 1″	9.6	♐	2° 53.9′	N 1° 39.1′
3	δ Ophiuci; Yed.		7m 18s	3.14	93° 22′ 10″	9.6		0° 33.3′	N 17° 15.8′
3	ε Ophiuci.		11m 43s	3.17	94° 23′ 12″	9.2		1° 45.6′	N 16° 27.3′
4	σ Scorpii.		13m 56s	3.63	115° 17′ 26″	9.0	♐	6° 3.3′	S 4° 1.1′
3–4	γ Herculis.		16m 24s	2.64	70° 33′ 8″	8.8	♏	2° 27.9′	N 40° 1.2′
1	α Scorpii (Cor.); Antares.		21m 53s	3.67	116° 9′ 10″	8.4	♐	1.1′	S 4° 33.1′
3	η Draconis.		22m 19s	0.82	28° 12′ 8″	8.2	♎	12° 40.1′	N 78° 27.5′
4	τ Scorpii.		23m 17s	3.90	124° 25′ 54″	8.3	♐	6° 43.7′	S 12° 39.7′
2–3	β Herculis; Korneforos.		24m 51s	2.58	68° 14′ 10″	8.2	♏	29° 18.6′	N 41° 51.7′
3–4	τ Scorpio.		28m 6s	3.72	117° 52′ 15″	7.8	♐	9° 12.8′	S 6° 6.1′
3–4	ζ Ophiuci.		30m 17s	3.30	100° 18′ 13″	7.6		7° 26.0′	N 41° 24.5′
2	α Triangulis Australis.		35m 27s	6.29	158° 17′ 40″	7.3	♐	10° 9.0′	S 46° 7.9′
3	ζ Herculis.		36m 35s	2.26	58° 10′ 10″	6.8	♏	29° 44.4′	N 57° 53.7′
3	η Herculis.		38m 37s	2.05	50° 50′ 20″	7.1	♏	27° 4.4′	N 60° 18.5′
3	ε Scorpii.		42m 1s	3.87	124° 3′ 58″	7.0	♐	13° 36.9′	S 11° 42.7′
3	μ¹Scorpii.		43m 24s	4.05	127° 19′ 51″	6.7	♐	14° 24.8′	S 15° 24.3′
4	μ²Scorpii.	16	43m 52s	+ 4.04	127° 48′ 10″	+ 6.6	♐	14° 30.1′	S 15° 21.8′

TABLE OF FIXED STARS.

MAG.	STAR.		RT. ASCEN.	ANN. VAR.	N. P. D.	ANN. VAR.		LONG.	LAT.
4–5	ζ¹ Scorpii.	XVI	45m 10s +	4.20s	132° 9' 11" +	6.6s	♐	15° 22.7'	S 19° 38.3'
3	ζ² Scorpii.		45m 47s	4.20	132° 8' 47"	6.8		15° 29.8'	S 19° 37.0'
4	ι Ophiuci.		48m 6s	2.84	79° 37' 39"	6.3		8° 53.9'	N 32° 31.6'
3	ε Herculis.		55m 30s	2.29	58° 53' 18"	5.6		6° 34.7'	N 53° 15.9'
3–4	η Scorpii.	XVII	3m 11s	4.27	133° 4' 18"	5.2		18° 59.6'	S 20° 9.3'
2–3	η Ophiuci.		3m 13s	3.45	105° 34' 4"	4.8	♐	16° 12.4'	N 7° 12.7'
3	ζ Draconis.		8m 26s	0.16	24° 7' 53"	4.5	♎	1° 26.5	N 84° 46.0'
3–4	α Herculis; Ras Algethi.		8m 57s	2.73	75° 27' 57"	4.5	♐	14° 15.9'	N 37° 17.0'
3–4	π Herculis.		10m 42s	2.09	53° 2' 54"	4.3		16° 19.1'	N 59° 33.4'
3–4	θ Ophiuci.		14m 20s	3.68	114° 52' 21"	4.0		19° 39.0'	S 1° 49.5'
3	γ Aræ.		14m 53s	5.03	146° 15' 24"	3.9		22° 32.9'	S 33° 5.5'
3	β Aræ.		14m 55s	4.07	145° 24' 33"	4.0		22° 25.4'	S 31° 54.9'
4	d Ophiuci.		19m 22s	3.82	119° 45' 8"	3.7		21° 8.1'	S 6° 36.7'
3	α Aræ.		22m 11s	4.02	139° 46' 27"	3.4		23° 11.4'	S 26° 32.4'
3–4	υ Scorpii.		22m 16s	4.07	127° 11' 32"	3.4		22° 16.1'	S 13° 59.2'
3	λ Scorpii; Lesath.		25m 8s	4.07	127° 0' 33"	3.0		22° 50.5'	S 13° 46.1'
2–3	β Draconis; Alwaid.		27m 36s	1.35	37° 36' 20"	2.9		10° 12.5'	N 75° 17.7'
3	θ Scorpii.		28m 20s	4.30	132° 55' 0"	2.9		23° 51.3'	S 19° 38.1'
2	α Ophiuci; Rasalague.		29m 8s	2.78	77° 20' 51"	3.0		20° 41.8'	N 35° 51.7'
3	κ Scorpii.		33m 51s	4.15	128° 57' 48"	2.4		24° 38.8'	S 15° 37.3'
3	β Ophiuci; Celbelrai.		37m 18s	2.96	85° 22' 43"	1.8		23° 35.7'	N 27° 57.1'
3–4	ι Scorpii.		38m 51s	4.20	130° 4' 39"	2.0		25° 46.7'	S 16° 41.8'
3–4	ξ Draconis; Grumium.		51m 22s	1.04	33° 6' 26"	0.7		23° 0.8'	N 80° 15.4'
2	γ Draconis; Etanin.		53m 42s	1.39	38° 29' 45"	0.6		26° 14.7'	N 74° 58.0'
4	67 Ophiuci.		54m 24s	3.01	87° 3' 39"	0.5		28° 26.2'	N 26° 24.0'
4	γ¹ Sagittarii.		57m 2s	3.84	119° 35' 3"	0.3		29° 21.2'	S 6° 7.4'
4	γ² Sagittarii.		57m 47s	3.86	120° 25' 25" +	0.4	♐	29° 31.1'	S 6° 57.8'
3–4	μ Sagittarii.	XVIII	6m 17s	3.59	111° 5' 21" —	0.6	♑	1° 29.1'	N 2° 21.9'
4	η Sagittarii.		9m 10s +	4.06	126° 47' 56"	0.6	♑	1° 53.3'	S 13° 20.7'
3	δ Ursæ Minoris.		11m 10s —	19.45	3° 23' 33"	1.1	Ⅱ	29° 27.3'	N 69° 55.8'
3–4	δ Sagittarii.		12m 59s +	3.84	119° 52' 45"	1.1	♑	2° 50.0'	S 6° 27.1'
3	ε Sagittarii; Kaus Aust.		15m 53s	3.99	124° 26' 30"	1.3		3° 20.1'	S 11° 1.8'
4	α Telescopii.		17m 42s	4.45	136° 2' 12"	1.3		3° 19.5'	S 22° 37.8'
4	λ Sagittarii.		20m 15s	3.71	115° 29' 20"	1.5		4° 34.4'	S 2° 6.6'
1	α Lyræ; Vega.		32m 42s	2.03	51° 19' 54"	3.1		11° 53.2'	N 61° 49.5'
3	β Lyræ; Shelick.		45m 28s	2.21	56° 46' 54"	3.8		17° 9.0'	N 56° 0.0'
3	σ Sagittarii.		47m 31s	3.73	116° 26' 39"	4.1		10° 38.2'	S 3° 25.8'
4–5	θ Serpentis; Alga.		50m 0s	2.98	85° 57' 26"	4.3		14° 0.0'	N 26° 53.6'
4	ξ² Sagittarii.		50m 16s	3.58	111° 16' 7"	4.4		11° 42.3'	N 1° 40.8'
3–4	ε Aquilæ.		53m 57s	2.72	75° 6' 2"	4.5		16° 31.5'	N 37° 35.2'
3	γ Lyræ; Sulaphat.		54m 16s	2.24	57° 28' 49"	4.6		20° 11.4'	N 55° 1.8'
3–4	ζ Sagittarii.		54m 39s	3.83	120° 3' 24"	4.7		11° 53.6'	S 7° 9.7'
4	τ Sagittarii.		59m 28s	3.75	117° 51' 2"	4.9		13° 3.6'	S 5° 3.7'
3	λ Aquilæ.		59m 37s	3.10	95° 4' 5"	5.1		15° 35.5'	N 17° 35.2'
3	ζ Aquil.; Deneb El Okab.18		59m 40s +	2.75	76° 19' 15" —	5.0	♑	18° 3.4'	N 36° 12.4'

ASTRONOMY.

MAG.	STAR.	RT. ASCEN.	ANN. VAR.	N. P. D.	ANN. VAR.	LONG.	LAT.
4–5	α Coronæ Australis. XIX	0m 58s +	4. 1s	125° 3' 48" —	5.1s	♑ 12° 23.2'	S 15° 17.5'
4–5	π Sagittarii.	2m 2*s	3.57	111° 19' 11"	5.4	♑ 14° 30.4'	N 1° 27.4'
3	ε Draconis.	12m 31s	0.04	22° 33' 31"	6.3	♑ 15° 32.6'	N 82° 53.0'
3–4	β¹ Sagitt.; *Urkab ur Rami*.	13m 36s	4.33	134° 41' 36"	6.1	♑ 14° 1.6'	S 22° 7.7'
4	β² Sagittarii.	14m 10s	4.33	135° 2' 4"	6.2	14° 4.7'	S 22° 28.7'
4	α Sagitt; *Rachbah ur Ramih*	15m 14s	4.18	130° 51' 1"	6.2	14° 53.3'	S 18° 21.4'
4	α Vulpeculæ et Anseris.	23m 30s	2.49	65° 35' 12"	7.0	27° 46.7'	N 45° 32.6'
3	β Cygni; *A'bireo*.	25m 41s	2.42	62° 18' 6"	7.3	29° 31.1'	N 45° 38.9'
4	α Sagittæ (The Arrow).	34m 31s	2.68	72° 16' 17"	8.0	29° 20.2'	N 38° 48.5'
3	γ Aquilæ; *Tarazed*.	40m 19s	2.86	79° 41' 24"	8.4	♑ 29° 12.0'	N 31° 15.5'
3–4	δ Cygni.	41m 4s	1.88	45° 10' 26"	8.5	♒ 14° 31.7'	N 64° 25.6'
1–2	α Aquilæ; *Altair*.	44m 41s	2.92	81° 22' 35"	9.1	0° 0.5'	N 29° 18.4'
3–4	β Aquilæ; *Alchain*.	49m 10s	2.95	83° 54' 18"	8.6	=	0° 41.2'	N 26° 41.4'
4–5	c Sagittarii.	54m 58s	3.70	118° 3' 18"	9.7	♑ 23° 19.1'	S 7° 3.5'
4	α¹ Capricor; *Dashabeh*. XX	10m 43s	3.33	102° 58' 35"	10.7	=	2° 1.5'	N 7° 0.2'
3	α² Capricor; *Secunda Giedi*.	11m 7s	3.34	102° 55' 51"	10.9		2° 6.7'	N 6° 36.6'
3–4	β Capricorni; *Dabih*.	13m 59s	3.28	105° 10' 28"	11.0	=	2° 18.1'	N 4° 36.6'
2	α Pavonis (Oculus).	15m 45s	4.79	147° 7' 38"	11.2	♑ 22° 4.1'	S 36° 14.8'
3	γ Cygni; *Sadr*.	17m 43s	2.15	50° 3' 33"	11.3		23° 7.3'	N 57° 8.0'
5	π Capricorni.	20m 10s	3.45	108° 37' 11"	11.5	=	2° 36.1'	N 0° 34.8'
3	α Indi.	28m 46s	4.25	137° 43' 31"	12.2	♑ 27° 21.1'	S 27° 44.5'
4	β Delphini; *Rotanem*.	31m 42s	2.81	75° 30' 19"	12.2	=	14° 36.1'	N 31° 55.9'
5	υ Capricorni.	32m 50s	3.43	108° 34' 37"	12.4	=	3° 55.1'	N 0° 14.2'
3	β Pavonis.	33m 36s	5.49	156° 39' 0"	12.4	♑ 20° 44.5'	S 45° 56.4'
3–4	α Delphini; *Svalocin*.	33m 50s	2.79	74° 31' 39"	12.4	=	13° 38.5'	N 33° 2.1'
1	α Cygni; *Deneb*.	37m 10s	2.04	45° 9' 56"	12.6	♓	3° 36.5'	N 59° 58.0'
4–5	ζ Capricorni.	38m 42s	3.57	115° 43' 8"	12.5	=	5° 25.0'	S 7° 0.6'
4–5	ε Aquarii.	40m 55s	3.26	99° 57' 8"	12.8		9° 58.7'	N 3° 5.6'
4	3 Aquarii.	41m 9s	3.17	95° 29' 3"	12.8		11° 13.4'	N 12° 23.5'
3	ε Cygni; *Gienah*.	41m 9s	2.43	56° 29' 50"	13.2		25° 39.4'	N 49° 25.3'
4–5	α Microscopii.	42m 10s	3.79	124° 14' 32"	12.6		3° 32.7'	S 15° 26.2'
5	χ Capricorni.	57m 17s	3.43	110° 20' 33"	13.9	=	10° 59.7'	S 2° 56.2'
5–6 61	Cygni. XXI	1m 18s	2.69	51° 51' 51"	17.5	♓	4° 56.6'	N 51° 52.3'
3	τ Cygni.	7m 37s	2.55	60° 17' 6"	14.5	♓	1° 18.8'	N 43° 42.2'
4–5	α Equulei.	9m 35s	3.00	85° 16' 5"	14.6		21° 26.6'	N 20° 6.2'
5	ε Capricorni.	15m 17s	3.36	107° 21' 55"	15.1	=	15° 56.1'	S 1° 21.4'
3	α Cephei; *Alderamin*.	15m 36s	1.44	27° 56' 39"	15.1	♈	11° 3.9'	N 68° 54.7'
3	γ Pavonis.	16m 5s	5.05	155° 55' 53"	15.8	♑	26° 49.4'	S 46° 59.2'
4	γ Capricorni.	19m 32s	3.44	112° 37' 5"	15.3		18° 11.4'	S 6° 38.9'
3	β Aquarii; *Sadalsud*.	24m 59s	3.17	96° 7' 12"	15.6	=	21° 39.1'	N 8° 37.5'
3	β Cephei; *Alphirk*.	27m 2s	0.80	10° 57' 17"	15.7	♉			8.5'
4	γ Capricorni.	33m 10s	3.34	107° 13' 33"	16.0	=			
2–3	ε Pegasi; *Enif*.	35m 3s	2.95	80° 41' 50"	16.3	♓			
4	θ Pegasi.	35m 59s	2.71	64° 55' 44"	16.3	♓			
3–4	δ Capricorni. 21	40m 5s +	3.3	106° 41' 36" —	16.1	=	21°		

TABLE OF FIXED STARS. 53

MAG.	STAR.	RT. ASCEN.	ANN. VAR.	N. P. D.	ANN. VAR.		LONG.	LAT.
3	γ Gruis. XXI	46m 22s	+ 3.67s	127° 57' 10"	— 16.58	♎	15° 30.8'	S 23° 2.3'
3	α Aquarii; *Sadalmelik*.	59m 22s	3.08	90° 55' 35"	17.3	♓	1° 36.6'	N 10° 10.1'
4–5	ι Aquarii.	59m 41s	3.25	104° 28' 31"	17.2	♒	26° 58.3'	S 2° 4.4'
2	α Gruis. XXII	0m 20s	3.81	137° 33' 35"	17.1		14° 9.6'	S 32° 53.6'
3	α Tucanæ (Am. Goose).	9m 55s	4.17	150° 52' 52"	17.7	♒	7° 55.1'	S 45° 23.4'
4–5	θ Aquarii.	10m 14s	3.17	98° 24' 18"	17.7	♓	1° 30.8'	N 2° 43.0'
3	γ Aquarii.	15m 12s	3.11	92° 0' 59"	18.0		4° 57.9'	N 8° 14.6'
4	ζ Aquarii.	22m 24s	3.09	90° 39' 31"	18.3	♓	7° 9.5'	N 8° 50.6'
4	β Piscis Australis.	24m 24s	3.14	122° 59' 14"	18.3	♒	25° 25.6'	S 21° 21.5'
4	α Lacertæ (The Lizard).	26m 9s	2.46	40° 21' 36"	18.3	♈	6° 24.9'	N 53° 17.1'
4	η Aquarii.	28m 56s	3.09	90° 45' 10"	18.1		8° 39.5'	N 8° 9.3'
3	β Gruis.	35m 12s	3.61	137° 32' 17"	18.6	♒	20° 33.9'	S 35° 25.2'
3	ζ Pegasi; *Homan*.	35m 14s	2.99	79° 49' 13"	18.7	♓	14° 24.6'	N 17° 4.1'
3	η Pegasi; *Matar*.	37m 9s	2.80	60° 25' 55"	18.7		23° 58.9'	N 35° 6.6'
4	μ Pegasi.	43m 58s	2.89	66° 3' 28"	18.9		22° 38.7'	N 29° 23.8'
4	λ Aquarii.	46m 5s	3.13	98° 11' 30"	19.0		9° 40.8'	S 0° 23.0'
3	δ Aquarii; *Scheat*.	48m 1s	3.19	106° 29' 5"	19.1		7° 7.7'	S 8° 11.1'
1	α Piscis Aust.; *Fomalhaut*.	50m 44s	3.33	120° 17' 3"	19.0		2° 5.9'	S 21° 7.0'
5	β Piscium.	57m 31s	3.06	86° 51' 0"	19.3		16° 50.6'	N 9° 3.3'
2	β Pegasi; *Scheat*.	57m 43s	2.90	62° 35' 40"	19.5		27° 37.6'	N 31° 9.5'
2	α Pegasi; *Markab*	58m 32s	2.98	75° 28' 0"	19.3		21° 44.7'	N 19° 24.5'
4–5	c² Aquarii. XXIII	2m 47s	3.21	111° 30' 59"	19.5		8° 16.9'	S 14° 29.1'
4–5	γ Piscium.	10m 41s	3.11	87° 24' 1"	19.6		19° 40.8'	N 7° 16.2'
4–5	ι Piscium.	33m 32s	3.11	85° 3' 4"	19.5		28° 53.7'	N 7° 10.5'
3	γ Cephei; *Er Rai*.	34m 14s	2.41	13° 3' 54"	20.1	♉	28° 21.1'	N 64° 38.9'
5	λ Piscium.	35m 10s	3.06	88° 54' 27"	19.8	♓	24° 51.1'	N 3° 25.2'
5	γ¹ Octantis.	44m 37s	3.62	175° 42' 50"	20.0	♑	10° 31.2'	S 65° 53.7'
4–5	ω Piscium.	52m 54s	3.08	83° 49' 43"	20.0	♈	0° 50.2'	N 6° 22.0'
4–5	30 Piscium.	55m 33s	3.08	99° 42' 31"	20.0	♓	26° 18.1'	S 5° 42.6'
4	2 Ceti. 23	57m 20s	+ 3.08	108° 1' 53"	— 20.1	♓	22° 1.0'	S 16° 14.1'

136. The annual variations in Right Ascension and North Polar distance, are themselves subject to a small change; that is, the variation is not the same in amount every year; it increases or decreases, as the star apparently approaches the pole or recedes from it. There are, however, only two stars among the 364 here catalogued, whose differences of variation are sufficiently great to require notice in ordinary calculations; they are both very near the North Pole (Sec. 45) in the constellation Ursa Minor; Alruccabah, the pole star, increases its annual variation yearly, at the rate of 0.1143 sec. in Right Ascension, and the Annual Variation in North Polar Distance of δ Ursæ Minoris changes at the rate of 0.028" yearly. Both these variations are additive for subsequent years, and subtractive for preceding years, to or from the places as found by the table — the correction here given being multiplied into half the square of the time in years before or after the epoch of the tables.

The declination of a star is readily known when the north polar distance is given. Thus:

90° — North Polar Dis. = the Dec; North.
North Polar Dis. — 90° = the Dec; South.

DEFINE AND EXPLAIN (the figures refer to the sections):
135. Table of Fixed Stars; Ann. Variation in Rt. Asc., N. P. D., and Lon. 136. Change in Ann. Variation; Declination.

CULMINATING, RISING, AND SETTING.

137. If from the Right Ascension of a star as given in the preceding table, (adding 24 hours, if required) we subtract the right ascension of the Sun at any required date (Sec. 34), the remainder will express the time that the Sun will pass the meridian above the earth, before the star, on the day in question; it is, therefore, the time after solar noon when the star culminates, or passes the meridian. If we correct this remainder by the equation of time (Sec. 34), it will give the hours and minutes past clock noon when the star culminates. Every star, the North Polar Distance of which is greater than the distance of the spectator from the North Pole of the earth, crosses the meridian South of the Zenith; hence the meridian passage, or culmination of a star, is often called "Southing."

138. To find the time of Rising and Setting of a star, necessitates a more lengthy operation. When the Sun is on the Equinoctial — near March 20th and September 23rd — the days and nights are equal all over the globe, excepting the increase always made in the length of the day by the refraction of the Sun's rays, which causes him to appear to rise earlier and set later, by a few minutes each day, than would be the case were not the solar rays bent downward a little in passing through the atmosphere. At the earth's equator he rises 6 hours and a few minutes before noon, and sets six hours and a few minutes after solar noon; but at any place between the Equator and the Poles, the difference between the length of the day and 12 hours increases with the Sun's declination, and with the same declination the difference increases with the latitude of the place of observation.

The Sun is longer above the earth than below it, at any given point on the earth's surface, when his declination is of the same name as the latitude of the place; and is longest below the horizon when his declination is different from the latitude. The United States are in North latitudes; therefore, when the Sun has North declination — from April to September — the days are longest; from October to March the Sun's declination is South, and the nights are longest (Sec. 17). The same difference is met with in noting the times of rising and setting of the fixed stars.

To an observer in North latitude 40°, the North Pole is always 40° above the horizon, and stars within 40° of the pole never sink below the horizon, being above it when they cross the lower meridian, as in the case of the Dipper (Secs. 36 and 40). The greater the North Polar Distance beyond this limit, the more time is occupied in the passage below the horizon; and stars, the North Polar distances of which are greater than 180° minus the latitude, are always below the horizon. A star on the Equinoctial is 12 hours above the earth, and 12 hours below it; the passage from the horizon to the meridian being measured by 6 hours, or (its equivalent) 90° of a great circle.

139. In the accompanying diagram let N S represent the horizon; P the North Pole, elevated above the horizon by an amount equal to the latitude of the place — say 40°; I, the South Pole; Z,

PASSING THE MERIDIAN AND HORIZON. 55

the Zenith; A B, the Equinoctial; E, the Eastern point of the horizon; T, the place of a star on the

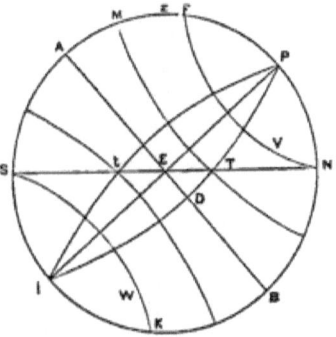

horizon. The angle E P A is equal to 90°; that is, a star at E would arrive at A in 6 hours afterwards. A meridian circle through the star at T, would cut the Equinoctial at D; and the Arc E D, measured by the angle E P D, is the excess over 90° of the angle A P D. In the spherical triangle E D T, we have given the angle at E (equal to 90° minus the latitude of the place), the right angle at D, and the side D T (equal to the declination of the star). Taking the side D E as the middle part, we have, by the Napierian analogy:

Cotangent of Angle E, multiplied into Tangent of D T, equals Sine of D E; or,

Log Tan of Latitude of Place, plus Log Tan of Star's Declination, equals the Log Sine of the Difference.

This difference, added to 90°, will give the arc D A, which, divided by 15 (Sec. 26), will give the hours and minutes required for a star to move from D to A, or from T to M.

It is apparent, from the diagram, that if the star were at *t*, with the same amount of declination South from A B, the difference above found would need to be subtracted from 90° to find the time of passage from the horizon to the meridian above the earth. It is manifest, also, that a star at V would be above the horizon during every part of its daily circuit; and a star at W, or at any point South of the circle S K, would never appear above the horizon.

140. The table on the next page is constructed in accordance with the above rule, and from it may be found, nearly, the difference in time between the horizon and meridian passage of any star. Look for the declination of the star in the first column, and find the Latitude of the place of observation at the top of the table. The difference will be found at the angle of meeting, proportion being made when the quantities used as arguments are intermediate to those given in the head and left hand column. Then:

141. If the place of observation be in North latitude, as is the case in the North American continent and Europe; the Semi-Diurnal Arc, or measure of the passage between the meridian and horizon, or horizon and meridian, is thus found :

If the Declination be North, add the tabular difference to 6 hours;

If the Declination be South, subtract the tabular difference from 6 hours;

The result is the Semi-Diurnal Arc. The rule reversed will give the arc for South latitudes.

Subtract the Semi-Diurnal Arc from the time of culmination, the remainder is the time of rising; add the arc to the time of culmination, the sum is the time of setting.

142. The arc and times thus found, make no allowance for refraction (Sec. 138) the effect of which is to make a star visible on the horizon while it is yet about 0° 35' below the line. This distance is, of course, measured perpendicularly from the horizon, and, as the stars rise and set more or less obliquely as the distance from the pole is less or

TABLE OF DIFFERENCES,

For finding the Ascensional Difference, from the Declination of a heavenly body and the Latitude of the place of observation.

	10°	20°	30°	35°	38°	40°	42°	44°	46°	48°	50°	52°	54°	56°	58°	60°
	h. m.	h. m.	h. m.	h. m.	h. m.	h. m.	h. m.	h. m.	h. m.	h. m.	h. m.	h. m.	h. m.	h. m.	h. m.	h. m.
2°	0 1	0 3	0 5	0 6	0 6	0 7	0 7	0 8	0 8	0 9	0 10	0 10	0 11	0 12	0 13	0 14
4°	0 3	0 6	0 9	0 11	0 12	0 13	0 14	0 15	0 17	0 18	0 19	0 21	0 22	0 24	0 26	0 28
6°	0 4	0 9	0 14	0 17	0 19	0 20	0 22	0 23	0 25	0 27	0 29	0 31	0 33	0 36	0 39	0 42
8°	0 6	0 12	0 19	0 23	0 25	0 27	0 29	0 31	0 33	0 36	0 39	0 41	0 45	0 48	0 52	0 56
10°	0 7	0 15	0 23	0 28	0 31	0 34	0 37	0 39	0 42	0 45	0 49	0 52	0 56	1 01	1 06	1 11
12°	0 9	0 18	0 28	0 34	0 38	0 41	0 44	0 47	0 51	0 55	0 59	1 03	1 08	1 13	1 19	1 26
14°	0 10	0 21	0 33	0 40	0 45	0 48	0 52	0 56	1 00	1 04	1 09	1 14	1 20	1 27	1 34	1 42
16°	0 12	0 24	0 38	0 46	0 52	0 56	1 00	1 04	1 09	1 14	1 20	1 26	1 32	1 41	1 49	1 59
18°	0 13	0 27	0 44	0 53	0 59	1 03	1 08	1 14	1 19	1 25	1 31	1 38	1 46	1 55	2 05	2 17
20°	0 15	0 30	0 49	0 59	1 06	1 11	1 17	1 22	1 30	1 36	1 43	1 51	2 00	2 11	2 22	2 36
22°	0 16	0 34	0 54	1 06	1 14	1 19	1 26	1 32	1 39	1 47	1 55	2 05	2 15	2 27	2 41	2 58
24°	0 18	0 37	1 00	1 13	1 21	1 28	1 35	1 42	1 50	1 59	2 08	2 19	2 31	2 45	3 02	3 22
26°	0 20	0 41	1 06	1 20	1 30	1 37	1 44	1 52	2 01	2 11	2 22	2 35	2 49	3 05	3 25	3 51
28°	0 21	0 45	1 12	1 27	1 38	1 46	1 55	2 04	2 14	2 25	2 37	2 52	3 08	3 28	3 53	4 28
30°	0 23	0 49	1 18	1 35	1 47	1 56	2 06	2 16	2 27	2 40	2 54	3 11	3 30	3 55	4 30	6 00
32°	0 25	0 53	1 25	1 44	1 57	2 06	2 17	2 29	2 41	2 56	3 13	3 32	3 57	4 32	6 00	
34°	0 27	0 57	1 32	1 53	2 06	2 18	2 30	2 43	2 57	3 14	3 34	3 59	4 33	6 00		
36°	0 29	1 01	1 39	2 02	2 18	2 30	2 43	2 58	3 15	3 35	4 00	4 31	6 00			
38°	0 32	1 06	1 47	2 13	2 30	2 44	2 59	3 16	3 36	4 01	4 34	6 00				
40°	0 34	1 11	1 56	2 24	2 44	2 59	3 16	3 37	4 01	4 35	6 00					
42°	0 37	1 17	2 05	2 36	2 59	3 16	3 37	4 02	4 35	6 00						
44°	0 39	1 22	2 16	2 50	3 16	3 37	4 02	4 35	6 00							
46°	0 42	1 29	2 27	3 06	3 36	4 01	4 35	6 00								
48°	0 45	1 35	2 40	3 24	4 01	4 35	6 00									
50°	0 49	1 43	2 54	3 46	4 34	6 00										
52°	0 52	1 51	3 11	4 15	6 00											
54°	0 56	2 00	3 30	4 58												
56°	1 01	2 11	3 55													
58°	1 06	2 23	4 30													
60°	1 11	2 36	6 00													
62°	1 17	2 53														
64°	1 25	3 13														

When the Sum of the Latitude and Declination exceeds 90° the Star neither rises nor sets; being always above the horizon or always below it.

greater, the time required for a star to rise through that perpendicular space varies with the latitude of the place of observation, and also with the declination. The following are the corrections in time due to horizontal refraction, for each 10° of latitude and declination:

CORRECTIONS FOR REFRACTION.

	0°	10°	20°	30°	40°	50°	60°	70°
	m. s.	m. s.	m. s.	m. s.	m. s.	m. s.	m. s.	m. s.
0°	2 20	2 22	2 30	2 42	3 03	3 38	4 40	6 41
10°	2 22	2 24	2 31	2 48	3 06	3 41	4 44	6 47
20°	2 30	2 31	2 38	2 52	3 14	3 52	4 58	
30°	2 42	2 48	2 52	3 07	3 31	4 12		
40°	3 03	3 06	3 14	3 31	3 59			

143. This correction always *increases* the semi-diurnal arc; it must, therefore, be subtracted from the already computed time of rising, and added to the time of setting, as found from the rule in Sec. 141. In the case of the fixed stars, the mere observer needs not to take this correction into account, as the stars are very seldom discernible within two or three degrees of the horizon.

In the case of the Sun, this correction being applied, will give the time when his *center* will appear to be on the horizon. If we wish to find the time when either edge of the Sun will touch the visible horizon, we must make a still further correction for his semi-diameter, which averages a little more than 0° 16', or about 46 hundredths of the perpendicular refraction. The tabular refraction being therefore augmented by nearly one half, will give the refractive correction for the Sun's upper edge; and diminished nearly one half, will be the refractive correction for his lower edge.

144. To find the length of the day from the preceding tables, requires a knowledge of the Sun's declination. It may be found from the table on the next page, which gives the right ascension and declination of every fifth degree of the Ecliptic, and for 5° of latitude North and South.

To find the Right Ascension of an object within 8° or 10° of the Ecliptic: Look for the nearest longitude in the fourth column of the Table. Then:

If the Longitude be in the first six signs, find the Right Ascension, without latitude, corresponding thereto, by proportion between the values in the second column.

If the object have north latitude, a comparison with the first column will show how much should be added or subtracted for 5° of latitude, and a proportionate quantity can be applied for the latitude given. If the object have South latitude, compare with the third column for the correction.

If the longitude be in the last six signs, find the right ascension without latitude, as before; compare with the third column for North latitude, and with the first column for South latitude. Then add 12 hours to the corrected quantity.

For Declination: Look for the longitude as before, in the fourth column; if the object be without latitude (as, the Sun), the declination will be found by proportion between the numbers in the sixth column; it will be North declination for the first six signs, and South for the last six.

If the object have North latitude, compare with the fifth column for the first six signs, and with the seventh column for the last six. For South latitude, compare with the seventh column for the first six signs, and with the fifth column for the last six. Where the declination in the table is marked —, the correction for 5° of latitude changes the declination to the other side of the Equinoctial, making it North instead of South, and *vice versa*.

145. *Example:* Find the times of culmination, rising, and setting of the Sun, Aldebaran, and Sirius, for May 20th, in North latitude 40°:

Sun's right ascension, May 15th is (page 12) 3h. 38m. 20s.
    do   do   June 1st, . . . 4h. 36m. 33s.

Proportionating: his right ascension, May 20th, is about 3h. 49m. 40s. Add to this the Equation

The following is a table for finding Right Ascension and Declination from Longitude and Latitude.

RIGHT ASCENSION.			LONGITUDE.	DECLINATION.		
♈ to ♎ 5° N. ♎ to ♈ 5° S.	ECLIPTIC.	♈ to ♎ 5° S. ♎ to ♈ 5° N.		♈ to ♎ 5° N. ♎ to ♈ 5° S.	ECLIPTIC.	♈ to ♎ 5° S. ♎ to ♈ 5° N.
h. m. s.	h. m. s.	h. m. s.				
23 52 4	0 0 0	0 7 56	♈ ♎ 0°	4 35′	0° 0	—4° 35′
0 10 24	0 18 21	0 26 17	5°	6 35	2 0	—2 36
0 28 41	0 36 45	0 44 39	10°	8 34	3 58	—0 38
0 47 20	0 55 14	1 2 59	15°	10 31	5 55	1 18
1 6 4	1 13 51	1 21 23	20°	12 26	7 50	3 10
1 25 3	1 32 38	1 39 59	25°	14 22	9 41	5 2
1 44 20	1 51 37	1 58 40	♉ ♏ 0°	16 8	11 29	6 48
2 3 48	2 10 51	2 17 36	5°	17 53	13 12	8 29
2 23 36	2 30 20	2 36 44	10°	19 32	14 50	10 5
2 43 52	2 50 7	2 56 6	15°	21 7	16 21	11 34
3 4 20	3 10 12	3 15 43	20°	22 34	17 46	12 56
3 25 14	3 30 34	3 35 32	25°	23 53	19 2	14 11
3 46 32	3 51 15	3 55 38	♊ ♐ 0°	25 2	20 10	15 16
4 8 10	4 12 13	4 15 57	5°	26 4	21 9	16 14
4 30 6	4 33 26	4 36 29	10°	26 54	21 58	17 2
4 52 20	4 54 52	4 57 12	15°	27 35	22 37	17 38
5 14 46	5 16 29	5 18 4	20°	28 5	23 5	18 5
5 37 18	5 38 12	5 39 0	25°	28 22	23 22	18 22
6 0 0	6 0 0	6 0 0	♋ ♑ 0°	28 27	23 27	18 27
6 22 42	6 21 48	6 21 0	5°	28 22	23 22	18 22
6 45 14	6 43 31	6 41 56	10°	28 5	23 5	18 5
7 7 40	7 5 8	7 2 48	15°	27 35	22 37	17 38
7 29 54	7 26 34	7 23 31	20°	26 54	21 58	17 2
7 51 50	7 47 47	7 44 3	25°	26 4	21 0	16 14
8 13 28	8 8 45	8 4 22	♌ ♒ 0°	25 2	20 10	15 16
8 34 46	8 29 26	8 24 28	5°	23 53	19 2	14 11
8 55 40	8 49 48	8 44 17	10°	22 34	17 46	12 56
9 16 8	9 9 53	9 3 54	15°	21 7	16 21	11 34
9 36 24	9 29 40	9 23 16	20°	19 32	14 50	10 5
9 56 12	9 49 9	9 42 24	25°	17 53	13 12	8 29
10 15 40	10 8 23	10 1 20	♍ ♓ 0°	16 8	11 29	6 48
10 34 57	10 27 22	10 20 1	5°	14 22	9 41	5 2
10 53 56	10 46 9	10 38 37	10°	12 26	7 50	3 10
11 12 40	11 4 46	10 57 1	15°	10 31	5 55	1 18
11 31 16	11 23 15	11 15 21	20°	8 34	3 58	—0 38
11 49 36	11 41 39	11 33 43	25°	6 35	2 0	—2 36
12 7 56	12 0 0	11 52 4	♎ ♈ 0°	4 35	0 0	—4 35

of Time (Sec. 34), which is 3m. 20s.; the sum is Sidereal, or Star Time, at noon of May 20th = 3h. 53m. — omitting the seconds.

Sidereal time being the right ascension on the meridian at clock noon, and the Sun's right ascension being less than that of the meridian, the Sun has passed the South line, 3m. 20s. at clock noon; that is, he culminates at 11h. 56m. 40s. A.M., on the day in question.

The Sun's declination is (Sec. 144) about 20° 5' North. The difference (Sec. 140) is 1h. 11¼m.; the refraction (Sec. 142) is 3¼m. The sum of these added to 6 hours, gives the semi-diurnal arc of the Sun's center = 7h. 14½m. This subtracted from 11h. 56½m. A.M., gives 4h. 42m. A.M. for the time of Sunrise, and added to 11h. 56½m., gives 7h. 11m. P.M. for the time of Sunset. Allowing 1¼m. (Sec. 143) for the Sun's semidiameter, we have :

Rising; Upper edge 4h. 40½m. A.M. — Center 4h. 42m. A.M. — Lower edge 4h. 43¼m. A.M.
Setting; Lower edge 7h. 9½m. P.M. — Center 7h. 11m. P.M. — Upper edge 7h. 12¼m. P.M.

From Table North Polar Distances	Aldebaran	Sirius	
Declinations (Sec. 136)	" 73° 45'	" 106° 33'	
Differences in Latitude 40° North	" 16° 15' North.	" 16° 33' South.	
Semi-diurnal Arc	" 0h. 57m.	" 0h. 58m.	
Refraction 3m., add, gives	" 6h. 57m.	" 5h. 02m.	
	" 7h. 00m.	" 5h. 05m.	
Right Ascensions	" 4h. 29m.	" 6h. 40m.	
Sidereal Time, Subtract	" 3h. 53m.	" 3h. 53m.	
Time of Culmination	" 0h. 36 P.M.	" 2h. 47 P.M.	
Corrected Semi-diurnal Arc	" 7h. 00m.	" 5h. 05m.	
Subtract; gives Rising	" 5h. 30 A.M.	" 9h. 42 A.M.	
Add; gives time of Setting	" 7h. 36 P.M.	" 7h. 52 P.M.	

146. The times of Rising and Setting are thus easily found to within a minute or two of the truth, which is near enough for most purposes. When, however, it is wished to be exact, the computation becomes much more voluminous. The Declination of the Sun, for instance, must be ascertained to the nearest second, both at rising and setting, because it is not the same at the two times, except when his Right Ascension is about 6 or 18 hours. The latitude of the place must also be ascertained accurately. Then, the refractive power of the atmosphere varies slightly with the height of the barometer, and the Sun's apparent diameter is not always exactly the same, being greatest when he is nearest to the earth — about the beginning of January — and least when the earth is in the opposite point of her orbit (See. 12). This book is not intended to teach these intricacies; but it is well that the student should know that accurate astronomical calculations involve the use of much time and an extensive acquaintance with mathematics.

DEFINE AND EXPLAIN (the figures refer to the sections):

137. To find the time of Culmination. 138. Rising and setting; refraction; effect of Sun's Declination; long days and short nights; near the South Pole. 139. Semi-diurnal Arc. 140. Differences. 141. How to apply tabular correction. 142. Difference due to Refraction. 143. Refraction for fixed stars; for the Sun; rising and setting of Sun's upper and lower limb. 144. Right Ascension and Declination from Longitude and Latitude. 145. Examples in Culmination; rising and setting. 146. Greater Exactness.

MAP NO. XIV.

147. This map shows, in diminished size, on a circular projection, all the prominent constellations visible in the United States. The center is the North Pole; the concentric circles are parallels of Declination, the third from the center being the Equinoctial; these parallels are those respectively of 30°, 60°, 90°, and 120°, of North Polar distance, the inside of the graduated circle representing 135° North Polar Distance, or 45° of South Declination. The eccentric circle is the Ecliptic, divided into twelve signs by as many portions of circles; it crosses the Equinoctial on the radial lines of 0h. and 12h. Right Ascension. Each of the smaller divisions on the outer circle corresponds to 1° in Right Ascension, or 4 minutes of time. The map shows all the Fixed Stars of not less than the third magnitude, with many of the smaller stars.

148. To find what stars are on the Meridian at any given time: Find the Sun's Right Ascension, or the Sidereal time, to the nearest minute, from the Table (page 12), and add to this the hours and minutes elapsed since the preceding noon, rejecting 24 hours from the total, if it exceed that amount. Look for the sum of the two times, on the outer circle of the map, and lay a flat ruler from this point across the map, passing through the center. The edge of the ruler will show the meridian line for the time assumed; that part of the line between the noted point and the center being the meridian above the pole; that portion lying on the other side of the center is a part of the meridian below the pole.

149. To find the position of the stars with respect to the horizon, at any time, in any latitude (North): From the diagram on the next page set a pair of compasses to the size indicated for any required latitude: Describe the circle on a piece of stiff, transparent paper, draw the meridian line, and mark on it the place of the North Pole. Cut out this circle, and lay it on the map, the Polar point coinciding with the center of the map, and the longer side of the meridian line coinciding with the upper meridian line found as in Sec. 148. The edge of the paper will show the place of the horizon; the straight line the meridian. The stars within the circle will be above the horizon; those without the circle will be below the horizon. Turn the paper round till the horizon line passes through any given star or constellation, and the difference of time, as pointed to by the meridian line, will show how much earlier or later than the assumed time the star, or group, will rise or set on that day.

150. To find the position of the Prime Vertical, or the circle which passes through the Zenith and the Eastern and Western Points of the horizon: Lay the circle on the map, the polar centers coinciding. Subtract the latitude from 90°, the remainder is the distance of the Pole from the Zenith. Estimate this distance from the Pole along the longer side of the meridian line, which can be done nearly enough on the 30° intervals. Draw an arc of a circle through that point and the two points marked "East" and "West" on the diagram. That curve will represent the prime vertical. It is represented on the diagram, for the latitude of 40°, by the curve marked "East" and "West."

151. The horizon circle may be drawn for any latitude between 20° and 60°, with sufficient exact-

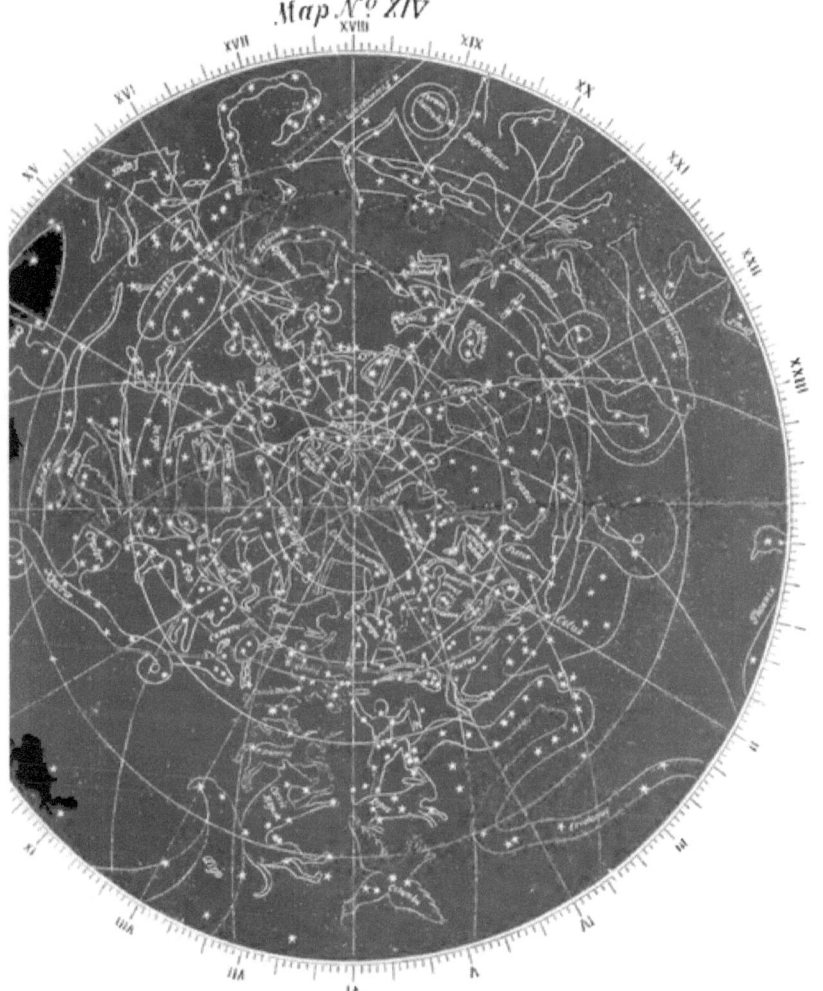

POSITION ABOVE THE HORIZON.

ness, by estimating for the intermediate position of the center, and striking the circle exactly through the points of intersection of all the circles in the diagram. Or, the circle of the horizon may be drawn independently, for any latitude, by finding in the table of differences (page 56) the difference for 30° or 60° of declination in the given latitude, setting off the semi-diurnal arcs from any meridian on Map No. XIV., on the Circles of Declination (parallels), and finding a point in the meridian from which, as a center, the horizon circle may be struck, passing through the given points on the Declination Parallels, and also crossing the Equinoctial in two exactly opposite points.

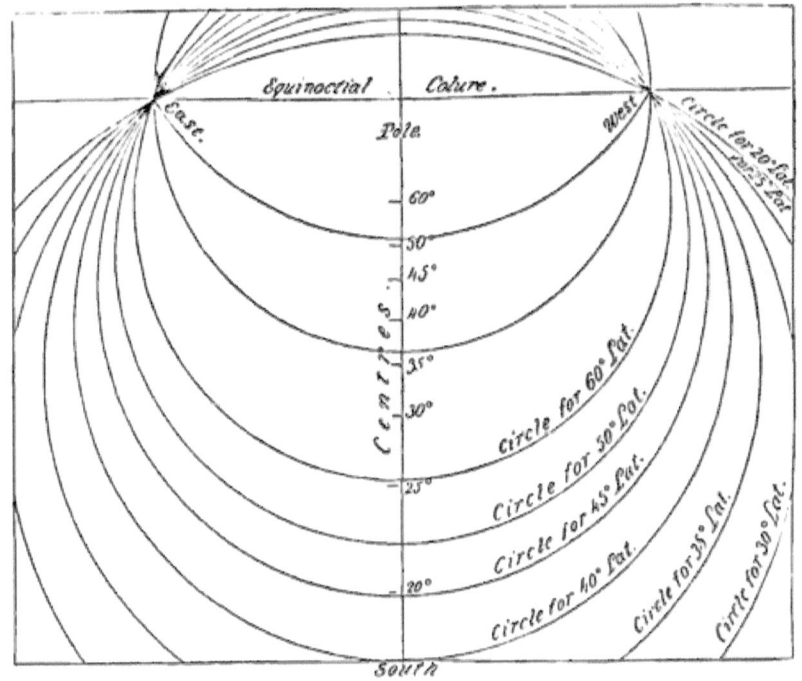

DEFINE AND EXPLAIN (the figures refer to the sections):

147. Map No. XIV.; concentric projection. 148. Stars on the meridian at any given time. 149. Position of Stars with reference to the horizon. 150. Prime Vertical. 151. Circle of the horizon.

THE SOLAR SYSTEM.

152. Several star-like bodies are often seen in the heavens, among the fixed stars already described, but changing their places with varying degrees of rapidity; always, however, within a few degrees of the Ecliptic line (Sec. 27). They sometimes appear to move forward with steady pace, increasing their longitudes nightly; then are stationary, keeping the same place among the fixed stars for several nights in succession; then seem to move backward among the stars, which is called Retrograding; and after another stationary period, again move forward.

The comparative prominence and brilliancy of each of these bodies varies with its rate of movement, the greater number being most brilliant when retrograding; but they all shine with a steady light, and by this peculiarity are easily distinguishable from the fixed stars, which twinkle. Viewed through the telescope, the difference is much greater; these bodies have then an appreciable, measurable, angular magnitude, while the brightest fixed star is but a luminous point. The reason of this difference has been ascertained to be, that these wandering bodies are comparatively near to us, and only shine with a reflected light, whereas the fixed stars are luminous bodies; shining, like the Sun, with their own light.

153. Five of these wanderers are so easily recognized, as to have attracted attention in a very early age of the world's written history. They are named Mercury, Venus, Mars, Jupiter, and Saturn. In common with our earth, more than a hundred smaller but named bodies, and, possibly, millions of others, still smaller and unnamed, they form a family called the Solar System. Each named member of the family, and probably every one, rotates on an axis of its own, in the same manner as does the Earth, and, like her, performs a circuit around the Sun. The time of revolution varies from eighty-eight days, in the case of Mercury, to nearly 165 years, in the case of the most distant known member of the family.

154. JUPITER.— The largest of these is Jupiter, a white-colored planet, with a bluish tinge; at times so brilliant that he has been said to cast a shadow — which, however, is an exaggeration. The table on the next page gives his approximate longitude and latitude during the last thirty years of the present century, for that time in each year when he is on the upper meridian at Solar midnight, and also at the maximum of his brightness for the year.

155. At these times the Earth is in direct line between the Sun and Jupiter, and the planet is said to be "in Opposition" to the Sun, rising when the Sun sets, setting when the Sun rises, and clearly visible in the Southeast on a fine evening. At the time of Opposition the planet is apparently Retrograding (Sec. 152), and continues to move backward for sixty days, at which time his longitude has decreased about 5°; that is: sixty days after Dec. 15th, 1870, the longitude of Jupiter will be II 16½°. At the expiration of this time he appears to remain stationary among the fixed stars for about five days, then commences to move forward to the East, preceding the Sun, at first slowly, then more

APPARENT MOTION OF JUPITER. 63

rapidly, but losing ground, so to speak, and setting a little earlier each evening, his brilliancy also gradually diminishing. Six months and a half from the time of Opposition, he is no longer visible to the naked eye; he is behind the Sun, and is then said to be "in Conjunction." A month later the planet is visible in the East just before Sunrise, and from then till four and a half months after the Conjunction, his distance from the Sun and his brilliancy increase, the planet moving forward among the stars, but so much more slowly than the Sun, that he is relatively receding, and rises earlier each morning. Then he becomes once more stationary for five days, and then retrogrades through 5°, in 57 to 60 days, when he is again in Opposition — on the meridian at midnight. The time occupied in this entire revolution, from Opposition to Opposition, is a little more than thirteen months, four of which have been occupied by the Retrograde movement, and nine by a forward motion in the Ecliptic. This is called the SYNODICAL REVOLUTION. During that time the planet has really moved forward a little more than 30° among the stars, and the Sun has passed over one annual circuit and about 30° more, just as the minute hand of a clock passes round the dial more than $1\frac{1}{12}$ times between two successive coincidences with the hour hand.

DATE.		LONGITUDE.	LATITUDE.	DATE.		LONGITUDE.	LATITUDE.
1869	Nov. 8	♉ 18°	S. 1° 18'	1886	March 20	♎ 1°	N. 1° 36'
1870	Dec. 13	♊ 21½	S. 0 29	87	April 20	♏ 1	N. 1 30
72	Jan. 15	♋ 25	N. 0 27	88	May 22	♐ 1½	N. 0 58
73	Feb. 15	♌ 26	N. 1 11	89	June 24	♑ 3	N. 0 9
74	March 15	♍ 26	N. 1 34	1890	July 29	♒ 7	S. 0 48
75	April 15	♎ 26½	N. 1 33	91	Sept. 5	♓ 13	S. 1 30
76	May 18	♏ 27	N. 1 4	92	Oct. 13	♈ 20	S. 1 37
77	June 20	♐ 29	N. 0 17	93	Nov. 18	♉ 26	S. 1 6
78	July 25	♒ 2½	S. 0 39	94	Dec. 22	♋ 1	S. 0 13
79	Aug. 31	♓ 8	S. 1 25	96	Jan. 24	♌ 4	N. 0 42
1880	Oct. 7	♈ 15	S. 1 39	97	Feb. 24	♍ 5	N. 1 21
81	Nov. 14	♉ 21½	S. 1 13	98	March 25	♎ 5	N. 1 37
82	Dec. 18	♊ 26½	S. 0 21	99	April 25	♏ 5	N. 1 27
84	Jan. 19	♋ 30	N. 0 34	1900	May 26	♐ 6	N. 0 53
85	Feb. 20	♌ 1	N. 1 16	01	June 29	♑ 8	N. 0 1

156. From the above table the place of Jupiter, near the Ecliptic, may be approximately found, almost by simple inspection, for any time during the remainder of the century. For two months before and after the time of opposition, he is within 5' of the place given in the table for that year, and during the other nine months of the thirteen he increases his longitude by an average of nearly 4' per month, or 1' per week. The latitude is never greater than that given in the table.

The Right Ascension and Declination may be known from the table on page 58, increasing or diminishing for the latitude; the times of culminating, rising, and setting may then be ascertained by the Rules in Secs. 137 and 141; or the times may be gathered by reference to the nearest fixed stars.

157. The phenomena above referred to are explainable only in one way. The Planet Jupiter moves around the sun once in about twelve years (from any star to the same star again, this being called his Sidereal Revolution), in an orbit exterior to that of the earth (Sec. 12). If the earth were the center of his motion, as was formerly supposed, the apparent movement would be regularly forward, as is that of the Sun; and the two hands of a clock (See. 155) would represent, also in uniformity, the relative motions of the Sun and Jupiter. We may, however, use the clock as an illustration, by supposing the hour hand to be lengthened to a little more than five times the length of the minute hand; Jupiter to be at its extremity, and the earth at the point of the minute hand, while the Sun is on the pivot in the center of the dial, the fixed stars being so remote that two lines drawn from any one of them to two opposite sides of the dial, would be parallel to each other.

The following diagram shows the relative positions at several times during one Synodical Revolution: The Arc A J represents the path of Jupiter between two conjunctions: B H is the space traveled by him in twelve months, divided into six equal portions of two months time, or about 5° of angular space. The complete circle represents the Earth's orbit, divided into six equal parts of 60°, each part corresponding to two months of her Annual Revolution. The center is the place of the Sun.

If the Earth and Jupiter be at the points marked A A, the sun will be in apparent conjunction with Jupiter; when, two weeks afterwards, they are at B B, the planet will appear to have moved forward from A to B, though the Sun will have moved over the greater angle measured by V E. Two months after the position B B, they will be at C C, the Sun being seen beyond F. Two months more brings the Earth and Jupiter to D D; and now the combined motion commences to change the direction of the line joining their centers, the Earth moving so much the more rapidly of the two, and the line F F points 10° backward from the direction of the line D D. The four months occupied in the passage from D to F is the period of Retrogradation; its middle point is the time of the opposition at E E, the Sun then being seen in the direction of B.

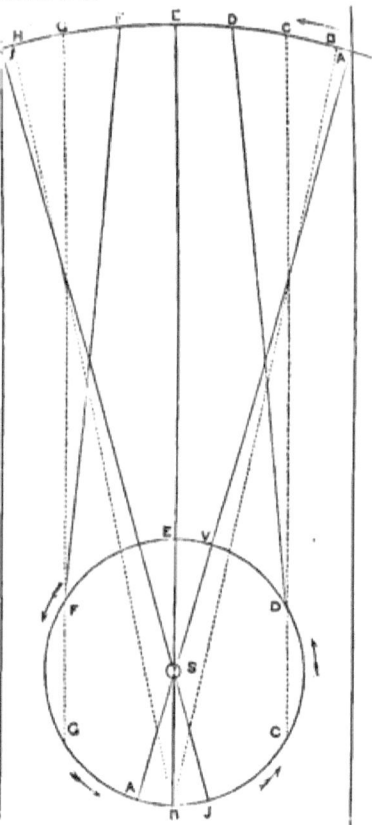

At F F the change of direction of the line ceases for a time, and Jupiter appears to be stationary, then to move forward in the direction of the arrows as the Earth passes round to J, where Jupiter is again seen as in Conjunction with the Sun. During the Synodic Revolution, the Earth has passed twice over the Arc A B J, equal to one month in excess of the twelve occupied by her own revolution. The Synodic period of the Earth and Jupiter is, more exactly, 398,867 days.

The position of a star, or planetary body, as seen from the Earth, is called its Geocentric position. The word "Heliocentric," as applied to place or motion, signifies that the Sun is assumed as the center of observation.

158. The latitude of Jupiter is accounted for by the fact that his orbit is not exactly coincident with that of the earth, but makes with it an angle of 1° 18' 42'',4., the two circles crossing each other in 9° 21' 27'' of the signs ♋ and ♑. The lack of equality in the yearly Arcs of the planet is due to the fact that neither of the two orbits is exactly circular (Sec. 16).

159. VENUS.— This planet, of a bright silver hue, is even a more prominent object in the heavens than Jupiter, but not so often visible, from the fact that she is never seen more than about 47° distant from the Sun, and is often so close to him that her brightness is darkened by the greater intensity of the solar rays.

Observation, sufficiently long continued, shows that Venus is visible from 45° to 47° East of the Sun, at intervals of a little more than nineteen months, being then seen in the West, soon after Sunset, and called THE EVENING STAR; from the time of this, her greatest elongation East, her distance gradually diminishes, till, in about seventy days, she is close to the Sun, and has, on two occasions, been seen to pass across the face of the luminary as a dark spot, showing that she is then *between* the Earth and Sun, and that the bright side of the planet is towards the Sun. This is called her Inferior Conjunction. From this time the Sun gradually leaves her behind, till, at the end of seventy days more, she is elongated 45° to 47° to the West of the Sun, rising before him in the East, and hence called THE MORNING STAR. She then again approaches the Sun, and in about seven months from the time of greatest elongation, she is again in Conjunction, but this time the Sun is between the Earth and Venus; hence this is called her *Superior* Conjunction. The planet is now moving at the rate of 1° 15' daily, and gaining on the Sun 0° 15' per day, a progress which soon carries her Eastward of the Sun into the position of Evening Star; she attains her greatest elongation in seven months from the time of the Superior Conjunction.

160. There is but one theory which enables us satisfactorily to account for these phenomena; it is, that Venus revolves around the Sun in an orbit within that of the Earth. The diagram on the next page shows the relative positions as above noted. The outer circle represents the Earth's orbit; the next to it is the orbit of Venus.

When the Earth and Venus are at A A, Venus is advancing directly towards the Earth, and therefore is not changing her apparent place among the fixed stars; this is the position at the time of her greatest Eastern Elongation, her apparent distance from the Sun being equal to the angle A' A' S. She is now the Evening Star. Fifty days afterwards, the Earth and Venus are at B B; this is the beginning of retrogradation (Sec. 152). Twenty-one days still later, they are at C C, and Venus is then in her Inferior Conjunction. At D D the retrograde motion ceases, and at E E the planet is at her greatest Western elongation. During the next seven months, while they are passing round to the position F F, Venus rises before the Sun, and is the Morning Star. At F F, she is in Superior Conjunction, and thence, for nine and a half months, till the position I I is reached, she is Evening Star — retrograding during the passage from H to K.

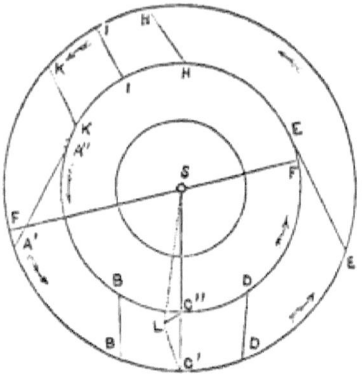

From the time of the Inferior Conjunction at C C to the one at I I, is a little more than one year and seven months; during that time the Earth has performed one revolution and seven-twelfths of another, and Venus has passed twice round her orbit, and described seven-twelfths of a third circuit. The mean time of the Synodical revolution is 583.92 days; the time of Venus' Sidereal revolution — as from C to C again — is 224⅔ days.

161. Venus is so much nearer to the Earth at the time of Inferior Conjunction than when they are on opposite sides of the Sun, that she would be about forty-five times brighter if she were equally luminous on all parts of her surface. But shining only with light received from the Sun on that side which is turned towards him, her apparent brightness does not materially vary, as near the time of her Inferior Conjunction her dark side is partially presented to us.

162. The following table shows the date and the longitude of Venus at the time of each Inferior Conjunction during the last thirty years of the present century. Her approximate position at any other time, may easily be found by allowing a proportionate distance, East or West. She retrogrades 8°, during 21 days, on each side of the Inferior Conjunction:

DATE.	LONGITUDE.	DATE.	LONGITUDE.	DATE.	LONGITUDE.
1870 Feb. 23	♓ 4½°	1881 May 1	♉ 11°	1892 July 5	♋ 15½°
71 Sept. 26	♎ 2½°	82 Dec. 6	♐ 16°	94 Feb. 14	♒ 27½°
73 May 4	♉ 11½°	84 July 8	♋ 17¾°	95 Sept. 17	♍ 25½°
74 Dec. 9	♐ 18½°	86 Feb. 17	♒ 30°	97 April 23	♉ 6°
76 July 11	♋ 20°	87 Sept. 20	♍ 28°	98 Nov. 30	♐ 11½°
78 Feb. 20	♓ 2½°	89 April 28	♉ 8½°	1900 July 2	♋ 13½°
79 Sept. 23	♎ 0¼°	90 Dec. 3	♐ 13½°		

163. Venus will cross the face of the Sun at the time of Inferior Conjunction, twice only during this period — in 1874 and 1882. At the other Conjunctions she will be North or South of the Sun, because the plane of her orbit does not exactly coincide with that of the Earth, but makes with it an angle of 3° 23′ 32″, cutting it in 15° 33′ 6″ of ♊ and ♐. While passing in her orbit from ♐ to ♊, her latitude is North, and from ♊ to ♐ it is South. When Venus is in Superior Conjunction, her apparent Latitude is of the same name as the Real Latitude, but less in amount, she being then much further from the Earth than from the center of her orbit at S. But at the Inferior Conjunction she is seen in the *opposite* sign to that occupied by her in reference to the Sun. Hence in February, 1870, being then more than 60° distant from the *nodes*, or points of intersection, her latitude in the orbit will be a little more than 3°, but it will be *North* latitude, she being then in ♍ as seen from the Sun. Her apparent latitude will also be much greater than the true — about 8½° — as, being nearer the Earth than to the Sun, her actual departure from the Ecliptic will be seen under a much greater angle than that subtended at the Sun; the distance C C is much less in the diagram than C' to S.

If L, in the preceding diagram, represent the position of the Ecliptic plane opposite C', then C' L will be the actual distance of Venus from the Ecliptic line, and the angle at S will be the latitude in her orbit, while the angle at C' will be the measure of her latitude as seen from the Earth — hence called the Geocentric Latitude, or the apparent latitude as seen from the Earth as a center.

164. MERCURY.— A dazzling bright speck of light, much smaller than Venus or Jupiter, is often visible in the East just before Sunrise, for a few days, and two months afterwards is seen in the West for three or four successive evenings, just after Sunset. This is the planet Mercury ; he makes an apparent swing, backwards and forwards across the place of the Sun, like a majestic pendulum, six of whose ticks nearly measure one of our Earth years. He is never seen more that 28° 20' from the Sun, and this distance alternates with about 16° as the limit of his greatest elongation on each side of the central luminary.

The inner circle of the preceding diagram represents the orbit of Mercury as relative to those of the Earth and Venus, except that his path is very far from being a perfect circle, he being much nearer to the Sun when in the middle of the sign ♊ than when in any other part of his orbit. A Sidereal revolution around the Sun, once in a little less than 88 days, gives a Synodical revolution of 115,877 days — nearly four months. The following are the approximate times of his Inferior Conjunctions during the last thirty years of the present century:

1869	June	24	1875	Oct.	29	1882	Feb.	22	1888	July	9		*Nov.	11
	Oct.	20	1876	Feb.	12		June	28		Nov.	1	1895	Feb.	23
1870	Feb.	2		June	16		Oct.	23	1889	Feb.	15		July	1
	June	4		Oct.	13	1883	Feb.	6		June	19		Oct.	26
	Oct.	3	1877	Jan.	27		June	7		Oct.	16	1896	Feb.	9
1871	Jan.	17		May	27		Oct.	6	1890	Jan.	30		June	10
	May	14		Sept.	27	1884	Jan.	20		May	30		Oct.	9
	Sept.	17	1878	Jan.	11		May	17		Sept.	30	1897	Jan.	23
1872	Jan.	1		*May	6		Sept.	20	1891	Jan.	14		May	20
	April	25		Sept.	9	1885	Jan.	4		*May	9		Sept.	23
	Aug.	31		Dec.	25		April	28		Sept.	12	1898	Jan.	7
	Dec.	16	1879	April	16		Sept.	3		Dec.	28		May	1
1873	April	6		Aug.	23		Dec.	19	1892	April	19		Sept.	6
	Aug.	13		Dec.	10	1886	April	9		Aug.	26		Dec.	22
	Dec.	1	1880	March	29		Aug.	16		Dec.	13	1899	April	12
1874	March	18		Aug.	5		Dec.	4	1893	April	1		Aug.	19
	July	25		Nov.	24	1887	March	21		Aug.	8		Dec.	7
	Nov.	14	1881	March	11		July	28		Nov.	27	1900	March	24
1875	March	1		July	17		Nov.	17	1894	March	14		July	31
	July	6		*Nov.	8	1888	March	3		July	20		Nov.	20

165. Mercury is retrograding, like Venus, at the time of Inferior Conjunction, and retrogrades during 11 or 12 days before and after, over an Arc of 4° or 5° on each side of the Conjunction. He attains his greatest Eastern elongation about 30 days before the Inferior Conjunction, and is then seen as an Evening Star; his greatest Western elongation is about 30 days after the Conjunction; he is then seen as a Morning Star. His Superior Conjunction is about equidistant in time between those of Elongation. The orbit of Mercury makes an angle of 7° 0' 20" with the Ecliptic, cutting it in 16° 50' 39" of ♂ and ♏. Hence the planet is generally North or South of the Sun at the time of Conjunction. He is not, however, seen from the Earth at any time with so great latitude as 7°, as, being much further from the Earth than from the center of the orbit, his actual departure from the plane of the Earth's motion is seen under a smaller angle than that subtended at the Sun (See. 163). He will cross the solar body, like a dark spot on the surface, at the four times marked with an * in the accompanying table. Mercury is the nearest to the Sun of any planetary body yet known.

166. MARS. — This planet is of a ruddy, sparkling hue, and very variable in his apparent size. His bright side is always turned, more or less, towards us, showing that he is never between the Earth and Sun (See. 161), while the great variation in his apparent magnitude shows that he is comparatively very near the Earth at the time of Opposition, and relatively far removed at the time of Conjunction. These conditions are satisfied by the theory that Mars revolves around the Sun in an orbit exterior to that of the Earth, but very much nearer than Jupiter. The following table shows the longitude and latitude of the planet, with the date, at each Solar Opposition during the last 30 years of the present century:

DATE.		LONG.		LAT.		DATE.		LONG.		LAT.
1869	Feb. 13	♌	23°	N.	4½°	1886	March 7	♍	16°	N. 4¼°
71	March 18	♍	29½°	N.	3¾°	88	April 11	♎	22°	N. 2½°
73	April 26	♏	6½°	N.	1¾°	90	May 27	♐	0°	S. 1½°
75	June 18	♐	27°	S.	3¼°	92	Aug. 3	♒	11°	S. 6½°
77	Sept. 2	♓	12°	S.	5½°	94	Oct. 20	♈	27°	S. 2½°
79	Nov. 11	♉	20°	S.	0½°	96	Dec. 11	♊	20°	N. 2½°
81	Dec. 27	♋	6°	N.	3½°	99	Jan. 19	♋	29°	N. 4½°
84	Feb. 1	♌	12°	N.	4½°	1901	Feb. 21	♍	3°	N. 4½°

167. Mars retrogrades through 5° to 10°, in 30 to 40 days before, and the same after, the Opposition, and is apparently stationary during two or three days before and after retrogradation. He is in Conjunction with the Sun about midway between the times given in the above table. His place at any other time may be found sufficiently near for purposes of general observation by remembering that, including the space lost in retrograding, Mars goes once round the heavens, and about 33° more, in the year and 11 months between his two stationary periods — or an average of about 17° per month.

Tracing him round the heavens in the same way as indicated in the case of Venus and Jupiter, we find that Mars moves in an orbit whose average diameter is 1½ times that of the earth; his Synodical revolution of 779.936 days resulting from a Sidereal revolution of 686.979 days, or nearly two years. His orbit is inclined only 1° 51' 6" to the plane of the Ecliptic, cutting it in 18° 33' 16" of ♂ and ♏, but he is sometimes seen with much greater apparent latitude when near the Opposition; he is then much nearer to the Earth than to the Sun, and his actual distance from the Ecliptic is therefore seen under a greater angle (See. 163). His greatest observed North Latitude is 4° 31′; the maximum of his South Latitude is 6° 47′; the difference is due to the fact that the relative eccentricity of his orbit brings him much nearer to the Earth when he is South of the Ecliptic, than is possible in the Northern half of his circuit.

168. SATURN. — This planet is of a pale, ashen color; he is not so prominent an object as either of those previously noted, though easily recognized. His motion among the fixed stars is comparatively regular, and so slow that his position changes but little for several days in succession. He makes a Synodical Revolution in 378,09 days, in which time he has moved forward but about 12°, making the Sidereal circuit in 10,759,22 days, or about 29½ years. He retrogrades 3½° during about 75 days before, and the same after, the time of the Opposition, and is stationary five days before and after Retrogradation.

The orbit of Saturn is exterior to that of Jupiter, the diagram for which will answer in the case of Saturn, by noting that the Earth is relatively but about half as far from the Sun as in the case of Jupiter. His greatest observed latitude is 2° 59′. His orbit makes an angle of 2° 29′ 24″ with the plane of the Ecliptic, cutting it in 22° [illegible]′ 37″ of ♋ and ♑. The following are the longitudes and latitudes of Saturn at the time of each Opposition during the last 30 years of the present century:

DATE.			LONG.		LAT.	
1869	June	4	♐	14 °	N.	1° 47′
70	"	16		25 °	N.	1° 20′
71	"	27	♑	6¾°	N.	0° 50′
72	July	9		18 °	N.	0° 18′
73	"	21		29½°	S.	0° 16′
74	Aug.	2	♒	11 °	S.	0° 49′
75	"	14		22¾°	S.	1° 20′
76	"	27	♓	4¾	S.	1° 49′
77	Sept.	8		16¼	S.	2° 13′
78	"	22		29¼°	S.	2° 32′
79	Oct.	4	♈	12 °	S.	2° 41′
80	"	18		25¼°	S.	2° 48′
81	"	31	♉	8¾°	S.	2° 43′
82	Nov.	17		24½	S.	2° 28′
83	"	27	♊	6¼	S.	2° 5′
84	Dec.	11		20½°	S.	1° 33′
1885	Dec.	25	♋	4¾°	S.	0° 56′
87	Jan.	8		18¾°	S.	0° 16′
88	"	21	♌	3 °	N.	0° 27′
89	Feb.	4		16¾°	N.	1° 4′
90	"	18	♍	0½°	N.	1° 39′
91	March	3		13¾°	N.	2° 7′
92	"	16		26¼°	N.	2° 28′
93	"	29	♎	9½°	N.	2° 42′
94	April	11		21¾°	N.	2° 46′
95	"	22	♏	4 °	N.	2° 43′
96	May	5		15¾°	N.	2° 35′
97	"	17		27½°	N.	2° 19′
98	"	29	♐	9 °	N.	1° 57′
99	June	10		20½°	N.	1° 33′
1900	"	22	♑	1¾°	N.	1° 3′
1901	July	4		13 °	N.	0° 32′

169. URANUS. — This planet is barely visible to the naked eye as a pale, bluish, star-like body, of the 6th magnitude. He is called, also, "Herschel," from the name of his discoverer, and "The Georgium Sidus," from the fact that he was discovered in the reign of George III. of England. Uranus is still more remote than Saturn, and changes his place among the stars much more slowly. His sidereal Revolution occupies 30,686.821 days, or about 84 years and six days; his Synodic Revolution is performed in 369.656 days, during which time he progresses but 4½° in longitude. He retrogrades 73 days, through about 2°, on each side of the Opposition. His orbit makes an angle of 0° 46′ 30″ with the plane of the Ecliptic, cutting it in 13° 17′ 9″ of ♊ and ♐. The following are the longitudes and latitudes at every fifth Opposition during the last 30 years of the present century. The place and date for intermediate years may be easily found by interpolation;

70   ASTRONOMY.

DATE.	LONG.	LAT.	DATE.	LONG.	LAT.
1870 Jan. 10	♎ 20°	N. 0 1°	1890 April 14	♎ 23°	N. 0 ¼
75 Feb. 2	♌ 15°	N. 0	95 May 8	♍ 17°	N. 0 ½
80 Feb. 24	♍ 6°	N. 0	1900 May 31	♐ 9°	N. 0°3'
85 March 26	♎ 0°	N. 0			

170. NEPTUNE. — This is the farthest removed from the Sun of any of the known planetary family, and is discernible only by telescopic aid. His Sidereal Revolution occupies 60,126.722 days, or 164 years and 7 months; his Synodical Revolution is 367.488 days, during which time he moves forward but about 2° in the Zodiac. He retrogrades 1°22' during 78 days, before and after the time of Opposition. His orbit makes an angle with the plane of the Ecliptic of 1° 47' 2", crossing it in 11° 9' 30" of ♌ and ♒. Neptune was discovered in 1846. The following is his longitude at each fifth Opposition from 1870 to the end of the century. His place for any intermediate Opposition may be easily found by interpolation:

1870 April 10	♈ 20°	1885 May 15	♉ 24°	1895 June 7	♊ 16°
75 April 22	♉ 2°	90 May 26	♊ 5°	1900 June 19	♊ 28°
80 May 3	♉ 13°				

171. The position of a planet among the stars at any time, may be readily found by the aid of the preceding tables, and reference to the maps in this book. Maps Nos. III. to X., inclusive, show the principal stars in all the Zodiacal constellations. The Ecliptic, near which the planets are always to be found, is divided on those maps into equal lengths of 15°, by circles of Latitude, and the position for intermediate degrees may be seen by inspection. Thus: On November 8th, 1869, the planet Jupiter is in opposition to the Sun, in ♉ 16°, and on December 13th, 1870, he is in opposition, in ♊ 21½. A reference to Map No. IV. will show that the first named position is near δ Arietis, and the second near ζ Tauri; the intermediate course of Jupiter may be traced among the fixed stars on Map No. IV., by reference to Sec. 156.

DEFINE AND EXPLAIN (the figures refer to the sections):

152. The Solar System; peculiarity of planetary appearance and motion; Retrograding; time of greatest brilliancy. 153. Planets known to the Ancients; the Solar family. 154. Jupiter; places at time of opposition. 155. Phenomena of opposition; retrograding; stationary; in Conjunction; Synodical Revolution. 156. Place of Jupiter at any time. 157. Motion round the Sun; Sidereal Revolution; exterior orbit; Synodic period. 158. Latitude. 159. Venus; greatest distance from the Sun; intervals of elongation; Morning and Evening Star; Inferior and Superior Conjunctions. 160. Inferior orbit. 161. Venus not self-luminous. 162. Times of her Inferior Conjunctions. 163. Transits of Venus; her Latitude. 164. Mercury; phenomena of elongation; Revolutions; Inferior Conjunctions. 165. Times of Retrograding and direct motion; inclination of orbit. 166. Mars; phenomena; how accounted for; dates of Solar Opposition. 167. Mars' latitude in the orbit. 168. Saturn, phenomena; slow motion, and causes; oppositions for 30 years. 169. Uranus. 170. Neptune. 171. Place of a Planet among the Fixed Stars.

THE MOON.

172. If we watch the course of the Moon for a few evenings, we find that she changes her place among the fixed stars much more rapidly than the Sun, or either of the planets, coming to the meridian nearly an hour later each day than on the day preceding: never retrograde, but with a very variable motion. Her period of describing the circuit of the heavens, or the time elapsing between two consecutive conjunctions with any fixed star, is easily found to be something more than 27 days; and from a mean of many hundred revolutions, it has been ascertained that her Sidereal Period is 27d. 7h. 43m. 5s.

The mean daily motion of the moon is therefore 13°10′36″, but a little observation enables us to see that her daily rate of movement is continually changing, it being sometimes less than 12°, at others more than 15°, in longitude. We also find, by comparison of anciently recorded observations with those of recent date, that the period of her revolution is slowly diminishing. During the time of her Sidereal Revolution, the Sun has progressed about 27° in longitude, and moves over about 2° more while she is passing that additional space. The Moon thus travels about 389° between any two consecutive Conjunctions — NEW MOONS — in 29.53059 days. This, the Synodic Revolution, is also called a LUNATION. Hence the Moon describes twelve complete Lunations in 11 days less than a year.

173. THE LUNATION.—That the Moon revolves around the Earth as the center of her motion, is evident from the fact that, unlike the planetary bodies (See. 152), she is never Retrograde. That she is not self-luminous, but receives her light from the Sun, is shown by the different phases she assumes, according to her position with respect to him.

The diagram on the next page will convey an idea of the principal phenomena of the Lunar motion. The inner circle represents the Earth; the outer circle is the Moon's orbit, or its projection on the plane of the Ecliptic. The Sun is supposed to be outside the limits of the drawing, his center in the direction of the line C S, beyond S; the other lines, running nearly in the same direction as C S, are supposed to be drawn to the outer edges of the Solar body or disc.

When the Moon is between the Earth and Sun, her illuminated side is turned directly from us; we see only a dark mass; she is then called "The New Moon." From this position she gradually increases in light, till at a distance of 90° in longitude, when we see one half of her Western, or illuminated surface, she then being East of the Sun, and setting after him; she is then in her First Quarter. As her angular distance from the Sun increases, we see more and more of her enlightened surface till she arrives at the "Full," when, being in Opposition, her illuminated side is turned directly towards us. From this, to the time of New Moon, her light diminishes, presenting the same changes, but in inverse order, till she becomes again totally obscured at her Conjunction with the Sun.

That the Moon is an opaque body is evident, from the fact that when passing directly between us and the Sun, or a star, she hides them completely from our view, producing, in the one case, an Eclipse; in the other, an Occultation.

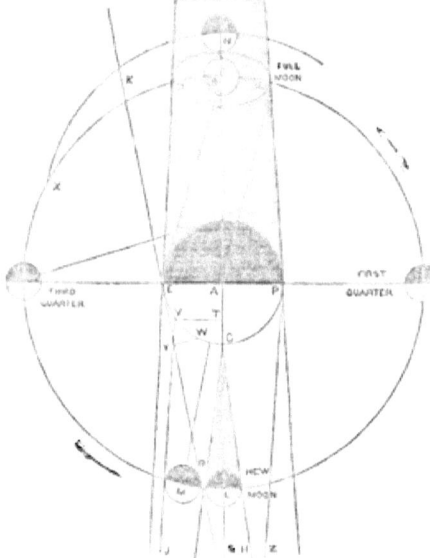

174. LATITUDE.—The Moon is seldom found on the Ecliptic. Her orbit makes an angle of 5° 8′ 48″, with the plane of the Sun's apparent path. The point where she crosses the Ecliptic into North Latitude is called the North Node, the place where she changes from North to South Latitude is the South Node. The places of the Lunar Nodes are constantly retrograding, making the circuit of the Ecliptic in about 18 years and 124 days, in consequence of which she performs her revolution from Node round to the same Node again, in about 27 days 5 hours; or in 2¼ hours less than the time of her Sidereal Revolution.

175. The Moon rotates on her own axis, but it has been found that her period of rotation is coincident with that of her Sidereal Revolution, so that she always presents the same side to us. Nevertheless, we see, at different times, much more than half of her surface; when she has great South Latitude her Northern side is more fully turned towards us, and *vice versa*. This balancing over, or vacillation, is called her Libration in Latitude. She has, also, a Libration in Longitude, her motion in her orbit being variable, while the rotation is uniform.

176. ECLIPSES.—When the centers of the Sun, Earth, and Moon are in one straight line, the light of one of the Luminaries is obscured for a time, and it is said to be Eclipsed. If this occur at the time of Full Moon, the lunar light is obscured during her passage through the Earth's shadow; if at the time of New Moon, the light of the Sun is prevented from reaching us by the interposition of the Moon, whose dark side is then turned towards us.

If the Moon's orbit coincided with the Ecliptic there would be an eclipse of the Moon at every Full, and an eclipse of the Sun at every New Moon. But her orbit being inclined to that of the Earth (Sec. 174), an Eclipse can only occur when, at the time of New or Full, the Moon is also in, or near, her Nodes; at all other times she will pass so far above or below the line B S as neither to suffer or cause privation of light. The greatest distances from the Nodes at which the Eclipse can possibly occur, are called the "Moon's Ecliptic Limits." In order to calculate these it is necessary to find, by observation, the angular magnitudes of the Earth, Sun, and Moon, as seen from each other.

177. PARALLAX.—The word means "change of place." A line from the Earth's center to the edge of the Sun's disc, nearly coinciding with the line C H, will make a small angle, at A, with the line A S, joining the centers of the Earth and Sun. This angle is the measure of the Sun's Semidiameter; it varies from about $0° 16' 18.2''$, in the beginning of January, to $0° 15' 46''$ in the beginning of July. The small angle formed at the Sun's center, by lines drawn from A and P, is the angular value of the Earth's Semidiameter, as seen from the Sun; it is about $8''.95$ — less than 9 seconds of Arc — and is called the SUN'S HORIZONTAL PARALLAX, it being the difference in the position of the Sun among the fixed stars, as seen at the same instant of time in the zenith of one observer, and on the horizon of another.

The Semidiameter of the Moon, as seen from the Earth's center, varies from $0° 16' 46''$ to $0° 11' 44''$. The angle subtended by the Earth's Equatorial Semidiameter, as seen from the Moon, is called "The Moon's Horizontal Parallax," being the angular difference of position of the Moon among the fixed stars, as seen at the same instant in the zenith of one observer, and on the horizon of another. This angle is about $1° 1' 25''$ at the Equator, when the Moon's apparent Semidiameter is greatest, and both diminish in the same proportion.

The change in the magnitude of these angles is evidently caused by a change in the actual distance between the respective bodies, as an object half a mile distant from the eye of the observer appears to be twice as large as if seen at the distance of one mile.

When the Moon is at her greatest distance from the Earth — her angular magnitude being least — she is said to be in *Apogee*; when at her least distance, she is said to be in *Perigee*. Similarly: the Earth's greatest distance from the Sun is called her *Aphelion*; her least distance is the *Perihelion*.

178. The Moon's angular magnitude may be measured very nearly by the aid of a simple apparatus. If we so fix a narrow board as that its flat surface will be turned towards the Full Moon, and move backwards or forwards till the Moon is apparently just shut out by the board from the view of one eye, the distance of the eye from that part of the board which is in a direct line with the Moon will be the radius of a circle, which, multiplied into 3.14159, and the product divided by 180, will give the measure of one degree to that radius (See. 22). Then half the width of the board, divided by one sixtieth part of the last found quantity, will give the number of minutes of space in the Moon's Semidiameter at the time and place of observation.

The Sun may be measured in the same way, by using a piece of smoked glass to prevent the eye from being dazzled by his rays. The method of finding the parallax is explained subsequently (See Secs. 194 and 197).

179. LUNAR ECLIPSES.—Let E A P, in the diagram, represent the horizon of the earth, and the base of her conical shadow, projected far beyond the Moon's orbit; F B D is a circular section of the shadow, at the distance of the Moon; A B, the distance of the centers of the Earth and Moon at the time of Full — the line A B being a continuation of that joining the centers of the Earth and Sun. We have given the angle A B E — equal to the Moon's Parallax, — and the angle made by the line E D with A B — equal to the Sun's Semidiameter diminished by his Parallax — to find F A B, or the angular Semidiameter of the shadow on the lunar orbit. The angle E B A, diminished by the angle formed by A B with E D, is equal to B E D. That is: Moon's Horizontal Parallax, minus Sun's Semidiameter, plus Sun's Parallax, equals D E B or B A F. This sum must, however, be increased by about one sixtieth part, for the refraction of the

Sun's rays in passing through our atmosphere. Let now X B represent a part of the Ecliptic; X, the place of the Moon's Node; and X N, a portion of her orbit; then, if at the time of Full, the edge — or limb — of the Moon, just touch the shadow without being obscured; it is evident that the latitude — B N — of the Moon's center, will be equal to the sum of the Semidiameters of the Moon and shadow. By Spherical Trigonometry, Tangent of B N, into Cotangent of X, equals Sine of B X — the distance from the Node beyond which an eclipse can not occur. B N is manifestly greatest when the Earth is in Aphelion, and the Moon in Perigee. That is:

Moon's Greatest Parallax	$+1°\ 1'25''$
Sun's least Semidiameter	$-0°15'46''$
Sun's Parallax	$+0°\ 0'\ 8\frac{1}{4}''$
Greatest Semidiameter of Shadow	$0°15'47\frac{1}{4}''$
Increase by one sixtieth part	$45\frac{1}{4}''$
Moon's greatest Semidiameter	$0°16'46''$
Maximum of B N	$1°\ 3'19''$
B N $=1°3'19''$; Log Tan	8,265,293
Angle X $=5°8'48''$; Log Cot	11,045,427
B X $=11°18'$; Log Sine	9,310,720

The angle at X is sometimes a little less than $5°8'48''$; hence the limit of Lunar Eclipses is extended to about 12° in longitude from the Node. If the Moon be in her Node without Latitude, as at B, the shadow being so much larger than the Lunar disc, the Eclipse is total, and of considerable duration.

If the distances of the luminaries be as above, the difference of diameters of Moon and shadow is $0°59'34''$, a space over which the Moon will pass in about 1h. 55m.: and for some time before and after this, a portion of the Moon's disc will be obscured, while she is entering and leaving the shadow. If the Moon be not in her Node, yet if the difference of Semidiameters be greater than her latitude, the Eclipse will be total, but not of so long duration; if the latitude be greater than the difference, the Eclipse will be but partial.

Besides this shadow proper — the *Umbra* — there is another — the *Penumbra* — external to the first, and surrounding it; the limit of this — E K — is determined by a line drawn from the opposite edge of the Solar disc; its Semidiameter is evidently equal to that of the Umbra, plus the Sun's diameter; it is not so dark as the Umbra.

An Eclipse of the Moon appears the same to all parts of the Earth's darkened hemisphere, because the Moon is really deprived of light while passing through the Umbra. She has usually a faint, dull hue, somewhat resembling that of tarnished copper — an appearance probably due to the Solar light refracted by the Earth's atmosphere.

180. SOLAR ECLIPSES. — Eclipses of the Sun occur only at the time of New Moon. In the diagram — having given the same quantities as for the Lunar Eclipse — the angle H C S will be the Sun's apparent Semidiameter; let M represent the center of the Moon. It is evident that the distance L M will be the limit of latitude of a Solar Eclipse, when the line K E H is a tangent to the surface of the Moon at M. When, therefore, the angular values are the greatest possible, we have:

L R = Sun's greatest Semidiameter	$0°16'18''$
R M = Moon's "	$+0°16'46''$
E R A = Moon's greatest Parallax	$+1°\ 1'25''$
A H E = Sun's Parallax	$+0°\ 0'\ 8''$
For Refraction of Moon's atmosphere add	$3''$
L M = Latitude Moon's Center	$1°34'40''$
L M $=1°34'40''$; Log Tan	8,440.027
Inclination $=5°8'48''$; Log Cot	11,045.427
Limit $=17°48\frac{1}{2}'$. Sine	9,485.454

The angle being sometimes a little less than $5°8'48''$ we assume 18° to be the limit of Solar

ECLIPSES. 75

Eclipses in longitude from the Node. If we now regard L M as a difference in longitude, we shall find that the Moon passes over the distance in 2h. 30m., nearly, which, doubled, gives 5 hours for the time of passage of the apex of the Umbra from E to P, — from the Western to the Eastern edge of the Earth's disc — when the Moon is in her Node. If the Moon be not on the Ecliptic, but still within limits, the shadow passes over a smaller portion of the Earth's surface, and the time of passage is lessened; if her latitude exceed the sum of the above quantities, her shadow passes Northward or Southward of the Earth, and the Sun is not eclipsed.

181. When the Earth is in Perihelion, and the Moon in Apogee (Sec. 177), the Sun's apparent semidiameter is 0'1'34" larger than that of the Moon (16'18" — 14'44"). If at that time the Moon's center be in a line with that of the Sun and the eye of the spectator, the dark body of the Moon is seen surrounded by a ring of light, the breadth of which is equal to the difference of the Semidiameters. This is called an *Annular Eclipse* — the apex of the shadow does not extend to the Earth's surface.

When the Earth is in Aphelion, and the Moon in Perigee, the Moon's Semidiameter is greatest by 0'1'0" only. The period of total darkness at any place can not, therefore, exceed the time required for the Moon to traverse that distance; the maximum time is 3m. 13s., but this is increased to the extent of a few seconds by the rotation of the Earth on her axis in the same direction. The mean rate of motion of the shadow over the Earth's surface is about 1,830 geographical miles per hour, or 30½ miles per minute.

The radius of the shadow at the Earth's surface will be nearly to the Earth's Radius, as the Tangent of the difference of Semidiameters is to the tangent of the Moon's Parallax. The Earth's radius being 3,962½ miles (Sec. 5), the radius of the shadow in the last case will be 64.7 miles. This is, approximatively, the half breadth of a zone, to every portion of which the Eclipse will be total. When exactness is required, a separate calculation must be instituted for every place from which it is wished to view the eclipse, as the variation of the Moon's Parallax during the time of Eclipse is often so great as to make a considerable difference in the result.

When the Sun and Moon are both at their mean distances from the Earth, the Lunar shadow just reaches the surface, and the space to which the Eclipse is total is reduced to a mathematical line.

182. The angular value — R — of the Penumbra — E R C — is equal to the sum of the semidiameters of the luminaries. Making L A radius, we have, L A into tangent of lunar parallax, equals A E, the Earth's radius. Similarly, L A into tangent of the penumbral angle, equals the semidiameter of the Penumbra at the Earth. That is, in the last case, the Penumbral limit is the Arc C V, or 31"38¾', equal to 0.5296 when the radius is 1. If the line of Central Eclipse be in the Zenith, as at C, this proportion gives the sine of the semispherical surface covered by the Penumbra, but in any other position — as M E — it expresses the perpendicular distance, in parts of the Earth's radius, of the limit from the axis of the Cone — as W Y.

The Sun is partially eclipsed at all places included in the Penumbra, but the times and phases are different for each place, as the apparent position of the Moon, with regard to the Sun, depends on her parallactic angle, which varies widely, at times, in places only a few miles distant.

In Solar Eclipses, the Moon's dark surface is usually marked by a faint light, which is probably due to reflection from the Earth's atmosphere; during the total Eclipse she is surrounded by a pale circlet, which has been attributed to the Sun's atmosphere, but is more probably due to a comparatively small lunar atmosphere.

183. The number of Eclipses in any year can not be less than two, nor greater than seven; the most usual number is four, and there are seldom more than six. If there be seven, five will be of the Sun; if only two, both must be Solar. There can never be more than three Lunar eclipses, and may be none at all. The reason of this is, that the Sun passes both Nodes but once in a year, unless he should pass one of them in the first few days of the year, in which case he will pass it again a few days before the year is finished, because the Nodes move backward (Sec. 174) about $19\frac{1}{3}°$ each twelve months, and the Sun passes from one to the other in about 173 days; if either Node be in advance of the Sun about 15° at the New Moon, he will be eclipsed, and at the subsequent Full the Moon will be eclipsed near the other Node, and will come round to the Sun's Conjunction, eclipsing him again about 15° from the Node; the same number may occur six months afterwards, near the other Node; when six more lunations are completed, and the Sun arrives again at the first Node, the year will lack but a day or two of being ended, and the Sun will be again eclipsed.

There may thus be three eclipses about each Node — two of the Sun, and one of the Moon — but if the Moon change close to either Node, she will be beyond the limits — 12° — at the preceding and succeeding Full, and in six months more she will change near the other Node, under the same circumstances, in which case there will be but one eclipse — of the Sun — at each Node.

184. In 223 mean lunations, the Sun, Moon, and Nodes return so nearly to the same relative positions, that the Node will be within 0°28'12" of its original place; at that interval, therefore, there will be a regular return of the same order of eclipses for many ages. The cycle is (Sec. 172) 29.53059×223; equal to 18 years, 11 days, 7 hours, 43 minutes, 3.648 seconds when Leap Year is only four times included, and 18 years, 10 days, 7 hours, 43 minutes, 3.648 seconds when Leap Year is included five times. In this period there are usually about seventy eclipses — twenty-nine of the Moon, and forty-one of the Sun; but although the number of Solar eclipses in this time is nearly as three to two of the Lunar, yet the Lunar eclipses visible at any one place are the most numerous, because the Lunar eclipse is visible to an entire hemisphere, while a Solar eclipse is only visible to a portion, and sometimes a very small part, of the Earth's enlightened side.

185. CHRONOLOGY. — Our ideas of time are dependent upon the apparent motions of the luminaries, and our methods of measuring it are solely referable to them. The year is measured by the Earth's annual motion round the Sun. This is completed in 0h. 11m. 12.43s. less than $365\frac{1}{4}$ days; we thus have three common years of 365 days each, and one Leap Year, containing 366 days. But these four years measure 44m.49.72s. more than the time of four revolutions, which excess amounts to 18h.40m. in a hundred years, and this excess is allowed for by omitting three leap years in every four centuries. Therefore:

To find the distance of any given year from Leap Year, we divide the number of years since the Christian Era by 4; if nothing remains, it is Leap Year; if 1, 2, or 3 remains from the division, it is so many years after Leap Year. But we count the even hundreds as common years, unless the number of centuries is exactly divisible by 4. Thus 1873, divided by 4, leaves 3 as the number of years after Leap Year; so 1700, 1800, and 1900 are common years, but 2000 will be counted as Leap Year, because 20 leaves no remainder after the division.

The commencement of the year is fixed as being

the time when the Earth is in Perihelion (Sec. 177), or as near thereto as is permitted by the even division of days into months. The subdivision of the year into months evidently corresponds to the nearest number of Lunations (Sec. 172) contained therein. They originally consisted of six months of 30 days each, and six of 31 days each; but when Augustus, the Roman Emperor, named the, then, sixth month, after himself, he made it a long month, and took the added day from February — then a short month, and the last; the year commencing with March. When it was subsequently found that the year was too long, the irregular month was chosen as the one to be further shortened; we now count 28 days in February, "Except in Leap Year, at which time, February days hath twenty-nine." In like manner, the Ecliptic is divided into twelve equal parts, or Signs, each being nearly the measure of the Sun's progress during one Lunation.

The Day is, of course, a natural division of time, being the period of the Earth's rotation: its length is uniform, as measured by any two successive returns of a fixed star to the same meridian, but variable as measured by the Sun (Sec. 16); it may be regarded as a year in miniature.

The subdivisions of a day are arbitrary, but their value is invariable; the number of hours from noon to midnight was probably suggested by the number of months in a year.

The Week — a collection of seven days — answers to the number of Planets and Luminaries known to the ancients; and, from the very earliest period of the world's history, the days of the week have each been designated by the name of one of those bodies. The week is also the nearest even number of days to the measure of the interval between two successive quarters of the Moon.

186. Fifty-two weeks of seven days make 364 days; hence, the same day of the month will fall one day later in the week each year, unless the 29th day of February intervenes, when it will fall two days later. Hence the following rule for finding the day of the week at any given date: Accounting the first day of January, in any year, as A, the 2nd as B...., the 7th as G, and the 8th as A again, the letter upon which Sunday will fall is called the DOMINICAL, or Sunday letter for that year. It is thus found:

From 1700 to 1800 take the year, from 1800 to 1900 take the year minus one, because 1800 was accounted a common year; for the same reason, from 1900 to 2000 take the year minus two; from 2000 to 2100 take the year minus two, because no additional day has been lost from the calendar. To the year, thus corrected, add its one-fourth part, omitting fractions; divide the sum by 7; the remainder, subtracted from 7, gives the number answering to the Dominical Letter, or the day in the first week in January on which Sunday falls.

Thus for 1875 we have $\frac{1875}{4} + 1875 - 1 = 2342$, which, divided by 7, leaves 4 remainder; this taken from 7 leaves 3, answering to C as the Sunday letter; or, Sunday falls on January 3rd, 1875, and, of course, on the 10th, 17th, 24th, and 31st of the month also. For the 1st of each month the letters are as follows: January and October, A; May, B; August, C; February, March, and November, D; June, E; September and December, F; April and July, G. Hence in 1875 the Sunday letter falls on the 1st of August, the 2nd of May, the 3rd of January and October, etc.; it must, however, be noted that in Leap Years the Dominical Letter, thus found, answers only for the first two months; for the rest, take the preceding letter. Thus, for 1872, the letters are G. and F.

187. A Solar Cycle is found by multiplying together 4 — the number of years from Leap to Leap — and 7 — the number of years in which the Dominical letter would run through the week, if there were no Leap Year. At the end of this period of 28 years, the days of the week fall on the same days of the month throughout the year; hence, if Sunday fall on the 2nd of January in 1870, it was the same in 1842, and will be so in 1898. The Solar Cycle began 9 years before the Christian Era; hence, to find the number of this Cycle, add 9 to the given year, and divide the sum by 28, the remainder is the number; thus for 1875 plus 9, the remainder is 8.

188. Nineteen Solar years contain 6939.603016 days, and 235 lunations are performed in 6939.-688=1565 days; that is, in 19 years the Sun and Moon return to the same relative positions on the same day of the month, and within one hour and a half of the same instant. This is called the METONIC Cycle, or *Golden Number*. It commenced one year before the Christian Era, therefore, if we add one to the given year, and divide the sum by 19, the remainder is the Golden number. But when a hundred year, not Leap Year, falls in this Cycle, the New and Full Moon will occur a day later than otherwise. The Golden Number for 1875 is 14.

By combining the Solar and Lunar Cycles, we obtain the Dionysian Period of 532 years, at the end of which time the New and Full Moon return to the same days of the week and month.

The Roman Indiction is a Cycle of 15 years, which began 3 years before the Christian Era. This is not an Astronomical Cycle, but is here mentioned as a component of the *Julian Period*, which is formed by the continual multiplication of these three Cycles — 28 × 19 × 15 = 7,980 years. The year 1800 was the 6,513th Julian Year: A. D. 1875 is Julian Year 6,588.

189. The *Epact* of any year is the Moon's Age, or the Number of days elapsed at the beginning of the year since the last New Moon. It is thus found: Multiply the Golden Number into 11, divide the product by 30; from the remainder (plus 30, if required) take 11; the remainder is the Epact. When the Solar and Lunar Cycles begin together, the Moon's age on the 1st of each month is as follows: January, 0; February, 2; March, 1; April, 2; May, 3; June, 4; July, 5; August, 6; September, 8; October, 8; November, 9; December, 10; consequently, if to the Epact we add the number above, and the day of the month, their sum, if under 30, is the Moon's Age; if above 30, the excess over 30 in months of 31 days, or the excess over 29, in months of 30 days, is the Moon's Age, or the number of days elapsed since the last preceding New Moon. Easter Sunday is the Sunday after the first Full Moon which occurs after the Vernal Equinox. Hence,

To find the date of Easter, in any year, we must remember that the Equinox falls on March 21st in common years, and on March 20th in Leap Year. Find the nearest New Moon to this date, and reckon forward 15 days to the Full. The Sunday next succeeding this is Easter.

DEFINE AND EXPLAIN (the figures refer to the sections):

172. The Moon; Sidereal period; mean daily motion; variable movement; Synodic Revolution. 173. The Lunation; Moon's Orbit; New and Full. 174. Moon's Latitude; Retrogradation of Nodes. 175. Rotation on Axis, Libration. 176. Eclipses, when; limits. 177. Parallax, angle at the centre; Angular Semidiameters of Earth and Sun; The Moon; Apogee and Perigee; Perihelion and Aphelion. 178. How to measure the Moon. 179. Lunar Eclipses; Earth's shadow; limit of distance from the Node; Umbra, Penumbra. 180. Solar Eclipses; limits; 181. Annular Eclipse; place and motion of shadow. 182. Penumbral limit. 183. Number of Eclipses in the year; the reason. 184. Lunar Cycle. 185. Chronology; Leap Year, Day, Week, and Month. 186. Place of Sunday; Dominical Letter. 187. Solar Cycle. 188. Metonic Cycle; Golden Number; Roman Indiction; Julian Period. 189. Epact, the Moon's Age; Easter.

ACTUAL DISTANCES AND VOLUMES.

190. The distances and dimensions hitherto dealt with, are angular only (Sec. 22). Two stars may be seen at the same angular distance, whether a few thousand or many millions of miles apart. We shall now indicate the character of the processes by means of which we are enabled, approximately, to measure the actual distances, sizes, and bulks, and comparative weights, of those bodies which are nearest to the Earth, and to fix certain limits of minima for those quantities in the case of the more distant. We pre-suppose a general acquaintance with plane and spherical Geometry, and plane Trigonometry.

191. THE EARTH. — The rotation of the Earth on her axis is uniform with regard to the Fixed Stars, and nearly so in relation to the Sun (See. 16). She presents every part of her Equatorial surface to the Sun once in about 24 hours. If we set a clock to the hour and minute of 12, at the exact instant of Noon at any place on the Equator, and then carry the clock Eastward till we arrive at some place where the same time movement shows one o'clock at the instant of Noon at the second place, it is evident that the two stations are one hour asunder, or one part in 24 of the Earth's circumference on the Equator. The distance between the two places being measured on the sea level, and multiplied into 24, the product is the perimeter of the Equatorial circle.

This operation, or a similar one, has actually been performed, and the circumference found to be 24,899 English miles; the diameter (7925.6 miles) is easily deduced from this by simple proportion; the half of the diameter, or radius is, more exactly, 20,923,599.98 English feet.

Similarly: If we proceed along a circle of the meridian (See. 7) from the Equator towards either pole, and, by means of the different altitudes of the Sun, or some fixed star, find the angle formed on that circle by lines pointing to the Zenith of each place, we shall arrive at nearly the same result; but upon taking accurate observations in several parts of the meridian circle, we shall find a slight inequality in the lengths of equal arcs in this direction; these quantities correspond to the different portions of the perimeter of an Ellipse, the longest axis of which lies in the plane of the Equator. The true figure of the Earth is hence that of an oblate spheroid, having its Polar diameter equal to 7899.1 miles. The Polar radius is 20,853,657.16 English feet.

192. This spheroidal shape is believed to be a consequence of the Earth's rotation on her axis. If the particles of matter of which she is composed were at some previous time in a fluid, or semi-fluid state, so as to allow a freedom of motion among them, the central tendency of the particles about the Equator would be partially counteracted by the whirling motion, or centrifugal force, generated there by the rotary movement, while the attractive force at the Poles, being undiminished, would press in, and partially heap up the more Equatorial portions; the comparatively small irregularities of the surface are traceable to other causes operating subsequently. Every star and planet sufficiently near, and large enough to be measured, is found to be similarly affected — the Equatorial diameter being the greatest.

193. A knowledge of the actual weight of the Earth is of little importance, and is probably unattainable. Nevertheless, an attempt has been made to solve the problem. A precipitous cliff was selected by Dr. Maskelyne, the internal composition of which was known, and its weight ascertained by multiplying its cubic contents in feet, into the weight of one foot of its material; the force of its attraction, as compared with that of the Earth's whole mass, was then tested by suspending a plumb line from its summit, and measuring the deflection from the perpendicular.

From these observations the weight of the Earth, as compared with that of an equal bulk of water, has been estimated as 5.4 to 1.0. It is very probable that the comparative weight, or specific gravity of the Earth, is not less than 5.4, nor greater than 6. It is thus possible to compute her weight in tons, true to within about ten per cent. of the gross amount.

194. THE MOON.—If two stations on the Earth's surface, several degrees apart, be selected, and the position of the Moon among the Fixed Stars be observed from each at the same instant, there will be found an appreciable difference. If the stations be so chosen that while the Moon is in the Zenith of one, she will be on the Horizon of the other, so that the line, A B, joining the centers of the Earth and Moon will form the longer leg of a right angled plane triangle, whose shorter leg is, A D, the Earth's Radius; then the angle A B D, at the Moon, will average about 0° 57' 20", and the angular Semidiameter of the Moon, the angle, B A C, will average 0° 15' 37.46", the proportion being as 3.6697 to 1 (Sec. 177): Hence by the theorem that "the sines of the angles are proportional to the opposite sides," we have

Sine of A B D : A D :: Cosine of A B D : A B.

Or Cotangent of B equals A B. Hence: Log Cotangent of 0° 57' 20" + Log of 3962.8 = Log of 237,626 miles; or, calling A D equal to 1, the result is 59.96435 Semidiameters of the Earth. In the same way we may find the Moon's distance from the Earth at Perigee and Apogee, by taking the greatest or least Parallax for the value of the angle at B (Sec. 177).

195. In the triangle A B C, we have known the side A B by the last proportion, and the angle at A equal to the Moon's angular semidiameter; whence the lineal diameter of the Moon is found to be 2153 miles. The solid contents of globes being proportional to the cubes of their respective diameters, the volume of the Moon is very nearly to that of the Earth, as the cube of 2153 to the cube of 7925.6; it is not exactly equal to the result thus found, because neither the Earth nor the Moon are perfect spheres, and the amount of departure from the true globular form is not precisely the same in each case.

196. It has been ascertained, by repeated observation, that at the time of the Moon's First Quarter, the Earth is about 0°0'6" in advance of the place in her orbit which she would occupy if the Moon were at the New or Full, and is the same angular distance behind her average place, if the Moon be in her last Quarter. That is: When the

Earth is preceded in her orbit by the Moon, she is retarded, and when the Moon is behind in the order of the Signs, the Earth is pushed forward by this amount. Astronomers have concluded from this fact that it is not the Earth's center, but a point between it and the Moon, which describes an equable circuit around the Sun.

In a subsequent section the distance of the Earth from the Sun, is stated as being a little over 91 millions of miles, whence this angular disturbance, which averages a little more than 0'0'.6", is found to correspond to about 2700 miles along the circumference of the Earth's orbit. The Earth's Equatorial Semidiameter of 3962.8 miles has an angular value (Sec. 177) of 8".95 in the orbit, whence, by proportion, the angular value of 2700 miles is a little more than 6". This distance from the center varies a little with the distance of the Moon from the Earth, but it is always in the position E, as in the diagram (Page 80). Around this common center of gravity the two bodies revolve each month, the point moving forward steadily in the Ecliptic.

The motion of the Earth is hence something like that of an eccentric wheel, while that of the Moon is similar to what it would be, with a shorter radius of revolution, if the Earth advanced equably.

As the Earth and Moon thus balance each other on this common center of gravity, it follows, from the laws of mechanics, that their forces are equal; that is: If the weight of each body be multiplied into its distance from the common center, the products will be equal. Dividing the Moon's distance by that of the Earth, we obtain about 87¾ for the number of times that the Earth is heavier than the Moon, whence the Moon's weight is 0.011399 when the Earth is taken as Unity.

If an average cubic mile of the Moon were of the same weight as an average cubic mile of the Earth, a comparison of their volumes would show the Moon's relative weight to be much greater than the amount here given. Comparing the relative weights with the relative bulks (Sec. 195), we find that the weight of the Moon is but 0.5657, where an equal bulk cut from the Earth would weigh 1. The last named fraction, therefore, represents the Moon's comparative density; and accepting 5.4 (Sec. 193) as the specific gravity of the Earth, the Moon is about 3 times as heavy as an equal bulk of water.

197. THE SUN. The Solar Parallax (Sec. 177), or the angle subtended at the Sun's center by two lines drawn to include the Semidiameter of the Earth, is 0'0'8".95; knowing the value of this angle, the distance of the Earth from the Sun can be calculated (Sec. 194) by multiplying the Cotangent of the parallax into the Earth's Semidiameter. But this angle is so small that it is very difficult to measure it, and an error of one-tenth part of a second in the observation would involve an error of a million of miles in the distance. Other means have, therefore, to be resorted to than those used in the case of the Moon.

Astronomers have had recourse to the motions of the planet Venus, for a determination of this angle from one much greater. The mean angular distance of Venus, at her greatest elongation from the Sun is 46° 20', which (Diagram, Page 66) is manifestly the time when she is in such a position in her orbit as to be at the right angle of a triangle, the hypothenuse of which is the Earth's distance from the Sun. Hence, by trigonometry, the sine of the angle of elongation, is her mean distance from the Sun, that of the Earth being taken as Unity. The natural Sine of 46° 20' is 0.7233317 nearly, which is the proportional distance of Venus from the Sun; the complement of this, or 0.2766683, is her proportional distance from the Earth at the time of Inferior Conjunction.

If, now, at the time of a transit of Venus over the Sun's disc (Sec. 163), observations be simultaneously taken at two different stations on the Earth's surface, the edge, C, of the planet will appear to the observer at A, as on the Sun's disc at D, and will appear at F as seen from B. The exact angular measure of the whole disc, N M, is already known, and that portion which lies between D and F is found by substracting therefrom the sum of the observed

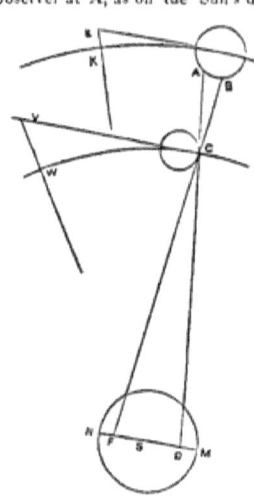

distances, N F and D M; the remainder, F D, when increased in proportion as C is nearer than A is to S, is the measure of the angle at C. The right line joining A and B is measurable, its extremities lying in the Earth's surface.

If at the time of transit, the line A B be equal to the Earth's radius, or 3962.8 miles, and the observed angular distance, D F, equal to 23′.4; this divided by C D, or 0,7233317, will give the angle at C equal to 32½′; the cotangent of half the last angle multiplied into half the base, will give the distance of the Earth from Venus as about 25,268,000 miles; whence, by proportion, the distance of Venus from the Sun is 66,060,000, and the Sum of the two gives 91,328,000 miles as the Earth's mean distance, from which, as Cotangent to the Earth's Radius, the Solar Parallax is given as 8″.95.

Observations similar to these were made at the times of transits of Venus in the years 1761 and 1769; the processes employed were similar to those here indicated, though not identical with them. The results arrived at were not precisely the same as now stated, but they have been corrected by subsequent observations of other bodies. Venus transits the Sun's disc twice during the present century — December 9th, 1874, and December 6th, 1882; the phenomena will be observed more accurately than in the last century, and it is believed that the results will not be materially different from those above given.

198. We can now calculate the diameter of the Sun, as we have already found that of the Moon (Sec. 194). His apparent semidiameter averages 0° 16′ 1″.82, the tangent of which angle is the lineal semidiameter of the Sun to the mean distance as radius. Thus:

Angular semidiameter = 0° 16′ 1″.82; Tangent is 7.668672
Mean distance (Sec. 197) 91,328,000; Log add 7.960604

Semidiameter = 425,868 miles;   Log is 5.629276

Twice this amount is 851,736 miles, the Sun's actual diameter.

The process of weighing the Sun is a much more delicate operation than in the case of the Moon, as a greater number of elements enter into the calculation. The principle is, however, the same, the Sun being weighed against the planetary bodies and the Earth, by noting the amount of their displacement in certain relative positions. His weight is found to be 354,936 times that of the Earth; and on comparing the two masses, we find that his density is but 0.284—that of the Earth being unity; or it is 1.333 as compared with an equal bulk of water as unity.

The distance of Venus, at the time of transit, being known (See. 197) we can calculate her diameter and bulk, as in the case of the Sun. These values, for all the principal members of the Solar family, are given in a subsequent table.

199. The mean angular semidiameter of the Sun has been used in this calculation, but this angular magnitude varies (See. 177) from $0°16'18''.2$ about the 2nd of January, to $0°15'46''$ about the 2nd of July. We conclude from this that the Earth is in Perihelion at the first mentioned time, and in Aphelion six months afterwards, as there can be no doubt that the actual magnitude of the Sun is the same at all times, and that the differences in apparent diameter are caused by our relative nearness or departure further from him. (See diagram on this page.) We conclude that the orbit of the Earth is not an exact circle, and that the Sun is not in its exact center.

The length of the Radius Vector — the distance from the Sun — at any point, being the cotangent of the angle of apparent semidiameter (See. 194), the Earth's radius being assumed as unity; the proportion between the cotangents of the greatest, mean, and least values, of the angle, gives the relative distances, or corresponding Radii Vectores, as 1.016751 for the Aphelion, and 0.983249 for the Perihelion, where the mean distance is 1. The difference is not large enough to be appreciable on a small scale. The following diagram shows, in exaggerated form, the difference in the angle under which the Sun's semidiameter is seen as viewed from E, the Aphelion, and C, the Perihelion:

200. EQUAL AREAS IN EQUAL TIMES. — The mean daily motion of the Earth in her orbit (See. 12) is $0°59'8''.33$; but this motion is not uniform, being most rapid at the Perihelion, and slowest at the Aphelion. We find by observation that this is the case with all the heavenly bodies, and that the rate of motion in the orbit bears a well defined proportion to the distance from the body around which the revolution is performed.

If the circumference of the orbit be divided into lengths, as E F, H K, D C, in the accompanying diagram, each of which is passed over in the same space of time, the product of the mean Radius Vector of each division into the lineal length of its corresponding arc, is a constant quantity. That is: $S L \times E F = S N \times H K = S M \times C D$; and because the area of a triangle with either a straight

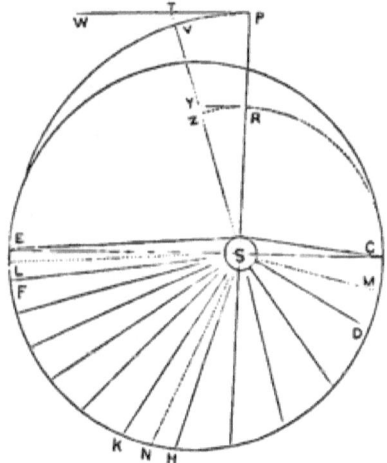

line or arc of a circle for its base, is half the product of the base into the height, therefore the areas of all the equal-timed triangles drawn in the orbit, as above, are equal to each other, and the planet, or Moon, describes equal areas in equal times.

201. If equal areas are described in equal times, then unequal areas are proportioned to the times of describing the arcs at the circumference, which form the bases of the triangles; and because similar plane figures have their areas proportional to the squares of their like dimensions, therefore similar triangles are proportional to the squares of the Radii Vectores.

In the two similar triangles S R Z, S P V, having equal arcs in different parts of the orbit, the times are proportional to the areas, and therefore to the squares of the Radii Vectores.

The force with which a body moves, being directly proportional to its velocity, is inversely proportional to the time, which increases as the velocity decreases. The relative force of motion at any point in the orbit is, therefore, inversely proportional to the square of the Radius Vector.

202. THE LAW OF ATTRACTION. — The universally accepted theory of the motion in the orbit is that it is the result of two forces, both acting in straight lines, at right angles to each other. A body once set in motion in the direction P to T, would move forward through the point W, and beyond it, in an unvarying straight line, if there were no other force operating; but we know that all masses of matter attract each other, as the Earth attracts a projected stone, and finally arrests its motion, only because the velocity of the stone is retarded in its passage by the atmosphere, and is at last overcome by the Earth's attraction.

A planet at P tends to move to T, but during the time due to describing the line P T, the attraction of the Sun has exercised a force equivalent to drawing the planet from T to V, and, in obedience to the two forces, the planet passes along the curve P V; the same forces constantly operating, the curve, P V, is many times reproduced till the planet has passed entirely round the Sun and is again at P. This second power we call the ATTRACTION OF GRAVITATION.

In order that a re-entrant curve shall be the result of the two forces, the attraction must always correspond in its variation of intensity to the force of motion in the orbit; that is: The force by which the body moves from P to V must vary exactly as the attraction which draws the planet from T to V; if it were otherwise, the attraction, being at any time the strongest, would continue so, and finally draw the planet to the center, while if the attraction were weakest the balance could not be recovered, and the planet would continually enlarge its orbit, till ultimately it would cease to revolve, and move off in a straight line. But the force of the motion in the orbit is (Sec. 201) inversely proportional to the square of the Radius Vector at any point, and the attraction must, therefore, at every point, be also inversely proportional to the square of the Radius Vector. In other words, the force of attraction increases as the square of the distance decreases.

The application of this law is universal. If we compute the Moon's orbit we find that her deflection from a right line — a tangent to the orbit — averages 0.053 of an inch per second of time, while a stone let fall near the Earth's surface will descend through a space of 193 inches in the same time. The distance of the Moon from the Earth's center (Sec. 194) is 59.96435 radii, or nearly 60 times as far distant as the Stone, and the distance — as TV — through which the Moon falls, is to the Stone's fall as the square of 1 to the square of 60, nearly.

203. RATIO OF DISTANCE. — If we compare the relative fall of two or more planets towards the Sun, we find the same law operating. The diagram on Page 82 represents portions of the orbits of the Earth and Venus. While the Earth is passing from A B to K, Venus describes the arc from C to W; the fall of the Earth from E to K bears the same proportion to V W, the fall of Venus in the same time, as the square of S W bears to S K; that is: The force of attraction to

LAWS OF PLANETARY MOTION.

the Sun is inversely proportional to the square of the distance from him. The proportion is easily verified by remembering that lines from S, to V and to E, are the Secants of the angles at the center, and that the Secant, diminished by Radius, gives the line of fall, which measures the force of the attraction.

Similarly: If we watch the planet Jupiter (See. 157) we find that the differences in his angular Semidiameter (Sec. 199) indicate that when in conjunction he is one-half more distant from the Earth than when in Opposition. That is: If when in Opposition the distance of Jupiter from the Earth be represented by 1, the distance is increased to 6 when in conjunction; but this difference is manifestly equal to twice the radius of the Earth's orbit, or Jupiter is 5.2 times farther from the Sun than is the Earth. The square of 5.2 is about 27, and, therefore, if the law hold good, Jupiter falls one foot towards the Sun, while the Earth falls 27 feet. Calculating his orbit on this basis, we find that it would carry him round the Sun in a little less than 12 years, and this result is confirmed by observation (Sec. 157).

If, now, we compare the times of Sidereal Revolution of the Earth, Venus, and Jupiter, with the relative distances as already found, we shall discover that they are related by a rather intricate proportion; the cube of the mean distance of each in miles, divided by the square of the number of days of Revolution, is equal in each case, and the same proportion is found to exist between every member of the system. The law is usually thus expressed: The squares of the Times are proportional to the cubes of the Distances.

204. By means of the preceding analogy, we can compute the actual distance from the Sun, of every member of the Solar system, having found its time of Sidereal Revolution by observation, and knowing the distance of the Earth or Venus (Sec.

197). From the apparent magnitude we can then calculate the actual diameter of each, while its comparative weight and density are ascertained by the more complicated process referred to in the case of the Sun (Sec. 198). The amount of displacement of any one member of the great family from its average path, can only be ascertained by an observation of all. The Sun, for instance, tends to oscillate around a point between it and the Earth (Sec. 196), but he also tends to oscillate around a point between himself and Jupiter, and the same with all the other planetary bodies. So when Venus is in line between the Earth and Sun, her attraction acts with that of the Sun, and draws the Earth some distance within its mean orbit, while the attraction of the Earth, acting against that of the Sun, but not so powerfully, draws Venus some distance outside of her average orbit; when Venus is in advance of the Earth, as at E E (Page 66), the mutual attraction retards the motion of Venus in her orbit, and accelerates that of the Earth, and the contrary effect is produced when the Earth is in advance, as at A A.

Every member of the system is thus eternally swaying back and forth, deviating first in one direction and then in another, in obedience to the ceaseless play of mutual attraction, but prevented from losing its own orbit by the original forward impulse. Several of these perturbations are sometimes acting in nearly the same direction, while at other times they counterbalance each other so accurately that no one is swayed out of the place it would occupy were there no disturbing influence.

205. NUTATION AND PRECESSION.— We are now prepared to comprehend, in part at least, the causes of the apparent irregularities in the Lunar motion (Sec. 172, and following). Reference to the diagram (Page 72) will show that the Moon moves absolutely over a greater space in her second and third quarters than the average motion, because she is then progressing faster than the Earth, while in the other two quarters it is less than the mean, as she is then falling behind the Earth though still going forward, because, in the two weeks from third to first quarter, the earth passes over a much greater space in her own orbit than is measured by the diameter of the Moon's orbit.

At the Full Moon the Sun and the Earth are both acting upon her in the same direction, shortening her Radius Vector, and still further accelerating her speed; at the New Moon the attraction of the Sun lengthens her Radius Vector, pulling her away from the Earth; and when in the third quarter the Sun hastens her motion, and retards it when she is in the first quarter, as he is then pulling her back partially in her course. If the Moon have latitude at the New or the Full, that is affected also by the Solar attraction.

In the next diagram the Sun is supposed to lie beyond S, the converging lines all meeting at his center; M represents the Earth's center, D the Moon at Full; I M D her latitude. The Solar attraction operates in the direction D to S, drawing the Moon to E, and increasing her apparent latitude to I M E. At the New Moon, C, the latitude is diminished to S M N, and a corresponding variation is necessarily produced in every other part of her orbit.

The Solar attraction thus causes variations in both the longitude, latitude, and distance of the Moon, aiding, or counteracting partially, the attraction of the Earth; and because the relative attraction is measured by the weight of the attracting body divided by the square of its distance (Sec. 202), a knowledge of the comparative weights and distances of the three is required to calculate the difference between the average — or mean — and the true place of the Moon at any time.

206. If the Earth were stationary, the places of the Lunar Perigee and Nodes (Sec. 174) would be always the same, coinciding with the place of the Full Moon; but in consequence of the Earth's Annual Motion, the Moon's Sidereal and Synodical Periods (Sec. 172) are not the same. Hence, the points of greatest attraction are continually changing — circulating round her orbit — and producing a revolution of the places of Perigee and Nodes, backwards through the signs.

207. The play of these mutual attractions (Sec. 204) affects not only the position of the Earth as a mass, but the position of the axis itself, or its inclination to the plane of the Ecliptic (Sec. 22).

Let V M R represent the Earth's greatest or Equatorial diameter. Then, of the protuberant mass about the Equator (Sec. 192) the portion at R will be more strongly attracted than the part at V, by the Sun beyond S, and the tendency of the attraction is to cause the two planes to coincide, bringing R V into the direction I S.

This force, however, is not constant, for V R represents the position only in midsummer and midwinter; at the Equinoxes the axis is perpendicular to I S, and the attractions are equal. Hence, the inequality increases from the Equinox to the Solstice, and thence diminishes gradually back to the time of the Equinox. The effect of this is a slight annual diminution in the angle between the Ecliptic and Equator, and also a vibration, or nodding, of the Earth's axis — hence called Nutation.

The Lunar attraction also operates, sometimes in the same direction with the Sun, as when at C or D; at other times partially counteracting his influence; and as the places of her Nodes change through the Signs, the situation of the greatest disturbing force varies also. Following the Lunar disturbing power, the Earth's axis describes a small circle of about 18° in diameter, in the same period as the Lunar Nodes (Sec. 174), and in the same direction — backward in the order of the Signs.

The swaying motion of the Moon (Sec. 196), also swings the Earth a little out of the plane of the Ecliptic when the Moon is in her greatest latitude, causing the Sun to appear to be not more than 4" on the other side of the Ecliptic, or average annual path of the Earth among the Fixed Stars as seen from the Sun.

208. The periods of the Luminaries being incommensurable — *i. e.* — no assignable number of revolutions of the one exactly measuring the period of the other, the result of their combined attraction on the Earth, is a gradual shifting of the points of intersection of the Ecliptic and Equatorial planes, as in the case of the Lunar Nodes, and in the same direction, but so much more slowly that 25,750 years are occupied in one revolution. The rate of change is about 50¼" annually. This movement is called The Precession of the Equinoxes (Sec. 25).

The period of 365 days, 5 hours, 48 minutes, 49 seconds, is the time of revolution from the Equinox round to the Equinox again — it is called a Tropical Year, as measuring the regular recurrence of the Seasons — but during this time the Equinox has receded 50¼", which distance has yet to be passed by the Earth ere she arrives at the original point, as referred to the Fixed Stars; this occupies about 20 minutes of time, so that we have another revolutional period called the Sidereal Year, of 365d 6h 9m, or more exactly, 365.2563714 Solar days (Sec. 157.)

209. THE TIDES. — The attractions of the Luminaries not only cause changes in the position of the Earth as a mass, but produce changes on her surface. The waters of the ocean are heaped up directly under the Moon's apparent path in her diurnal journey, the point of greatest elevation following her from West to East, and shifting back and forth across the Equator, as the Moon's declination varies, during each lunation. This is called the Tidal Wave.

In the open ocean the crest of the wave moves from East to West, on the circle of Latitude which corresponds to the Moon's declination, but about 46° degrees behind the line joining the centers of the Earth and Moon; the time of high tide being therefore about three hours (Sec. 26) later than the time of the Lunar culmination (Sec. 197). Directly opposite, at the antipodes of this wave, another protuberance of waters is met with. From each of these summits, the waters slope down in all directions, till at a distance of 90° from each there is a corresponding depression of the waters below the natural level.

210. These points of greatest protuberance and depression move round the Earth once in about 24 hours 53 minutes, that being the average interval of time between two successive returns of the Moon to the meridian of any place (Sec. 172). Hence at any place on the ocean, the two points of elevation succeed each other at mean intervals of 12h 26m, and the points of depression succeed at the same intervals, giving an average time of 6h 13m between the occurrence of high and low tide.

In the open ocean the direct high tide — the first mentioned — is 3 hours behind the Moon, but where its flow is obstructed by a continent, or retarded in the passage through narrow inlets, the interval between the meridian passage and the high tide is changed, so that each place on the sea coast has practically its own time of high tide.

211. The cause of the direct tide is found in the greater distance of the Moon from the Earth's center, than from the waters on the surface. The difference is equal to the Earth's radius (Sec. 191) — nearly 4,000 miles. The water is, therefore, attracted more strongly than the Earth (Sec. 203), and is partially heaped up, falling behind the perpendicular line in just the same way that a weight suspended by a long string, will hang back when the upper end of the string is carried rapidly forward. The opposite protuberance is usually ascribed to the greater attraction of the Moon for the Earth's center than for the waters of the ocean nearly 4000 miles further distant.

That the elevation of the waters is due to the Lunar attraction, is proven by the fact that the elevation varies with the distance of the Moon from the Earth. The comparative heights of the wave are nearly as follows: Moon at mean distance, 51; Moon in Perigee, 59; Moon in Apogee, 43.

212. The height of the tides varies also with the position of the Moon with reference to the Sun. The attractive power of the Sun for the whole Earth, as a mass, is 210.8 times that of the Moon, but the differences in his attractive power are much less, owing to his greater distance; that is, 91,328,-000 miles, compared with 91,328,000 minus 3963, shows a much less ratio than 237,626, compared with 237,626 minus 3963. The attraction of the Moon at her mean distance being represented by 51 (Sec. 211), that of the Sun will be equal to 20; at the Earth's Perihelion distance, 21; at the Aphelion, 19.

When the Sun and Moon are both acting in the same direction — as at the time of New or Full Moon — the tide rises higher, the mean height being then represented by 20 plus 51 (= 71), the sum of the numbers measuring the attractions of the Luminaries; this is called Spring Tide. At the time of the First and Third Quarters, the Sun is operating to elevate the depressions caused by the Moon, and the mean height of the tide is represented by the difference of the numbers: 51 minus 20 (= 31); this is called Neap Tide. If the Luminaries be both at their greatest distance from the Earth at the time of Neap Tide, the difference becomes (43 — 19) 24; if both be at their least distance at time of Spring Tide, the sum is (21 + 59) 80. The greatest tide is to the least, as 80 to 24; or as 10 to 3.

Where the tidal wave reaches a narrow bay or inlet through a wide channel, the tide rises to a greater height, the water rushing in much more rapidly than it can flow away. The average height of the tide decreases as the latitude of the place increases its distance from the crest of the wave.

DEFINE AND EXPLAIN (the figures refer to the sections):

190. Actual distances, bulks, and weights. 191. The Earth; Equatorial and Polar Diameters. 192. Spheroidal shape; inferences. 193. Earth's actual weight, as compared with water. 194. Mean distance of Moon from Earth. 195. Moon's diameter and volume. 196. Her weight and density. 197. Distance from the Sun; mode of finding Solar Parallax; transit of Venus; her distance. 198. Size and weight of Sun. 199. Perigee and Apogee; lengths of Radii Vectores. 200. Equal Areas in Equal times. 201. Proportion of time and force to square of Radius Vector. 202. The law of attraction; the Moon's fall per second, fall of a stone; their ratio. 203. The Earth, Venus, and Jupiter; comparative velocities and attractions to the Sun; the Squares of the Times proportional to the Cubes of the Distances. 204. Mutual attractions, and ceaseless perturbation; weight of the Planets. 205. Nutation and Precession; cause of irregularities in the Lunar Motion. 206. Revolution of Moon's Perigee and Nodes. 207. Nutation of the Earth's Axis; 208. Precession of the Equinoxes. Tropical and Sidereal years. 209. The Tides; position of Wave Crest. 210. Points of high and low tide; intervals. 211. Causes; ratio of force at Perigee and Apogee. 212. Sun's attraction; Spring and Neap Tides.

ELEMENTS OF THE SOLAR SYSTEM. 89

213. The following tables show the distances, magnitudes, and bulks, with other elements of the orbits, of the leading members in the great Solar family. The quantities are taken from the astronomical works of Watson, and Loomis, but are modified to correspond with the recently announced change in the estimated value of the Solar Parallax, and adapted to the beginning of the year 1875. The first table gives the elements of magnitude of the planetary orbits. The mean, greatest (Aphelion), and least (Perihelion), distances from the Sun are represented in comparison with the Earth's mean distance as Unity. The eccentricity of each planet's orbit is given in decimal parts of half the longest diameter of its own Orbit.

Name.	Distance from the Sun.			Eccentric'y.	Mean Distance in Miles.
	Mean.	Greatest.	Least.		
Mercury,	0.387098	0.466687	0.307510	0.205603	35,353,000
Venus,	0.723332	0.728261	0.718403	0.006814	66,060,010
Earth,	1.000000	1.016751	0.983249	0.016751	91,328,000
Mars,	1.523692	1.665964	1.381420	0.093374	139,156,000
Jupiter,	5.202776	5.453961	4.951591	0.048281	475,158,000
Saturn,	9.538786	10.072187	9.005385	0.055919	871,163,000
Uranus,	19.182390	20.077455	18.287325	0.046661	1,751,893,000
Neptune,	30.070552	30.326040	29.815064	0.008496	2,746,800,000

The following table gives the elements of Motion; the Sidereal Revolution, or the time in days, and decimal parts of a day, occupied in making the circuit from any star, round to the same star again; the Synodical Revolution, or the interval between any two successive conjunctions with the Sun as seen from the Earth; the average motion in the orbit during each earth-day, the length of the day of each body, and the ratio of difference between the Equatorial and Polar diameters.

Name.	Sidereal Revolution.	Synodical Revolution.	Mean Daily Motion.	Rotation on Axis.	Compression.
Mercury,	87.969282	115.877	245′ 32.6″	24h 5m 28s	1/5
Venus,	224.700775	583.920	96′ 7.8″	23h 21m 21s	
Earth,	365.256374	......	59′ 8.3″	23h 56m 4s	1/315
Mars,	686.979456	779.936	31′ 26.7″	24h 37m 22s	2/75
Jupiter,	4,332.581803	398.867	4′ 59.3″	9h 55m 26s	1/17
Saturn,	10,759.219711	378.090	2′ 0.6″	10h 29m 17s	1/12
Uranus,	30,686.820556	369.656	42.4″	9h 30m ?	1/9
Neptune,	60,126.722000	367.488	21.6″	not known.	
Sun,	days.	days.		607h 48m	

The following table shows the relations of position in the orbit; the longitude of the planet when nearest to the Sun (the longitude of the Aphelion is in the same degree and minute of the Opposite sign); and the point where the planet crosses the Ecliptic from South to North Latitude (the South Node

ASTRONOMY.

is in the same degree and minute of the **Opposite Sign**); the third and fourth columns give the annual movement forwards (+) or backwards (—) along the Ecliptic, as referred to the Fixed Stars; in addition to these variations, the Longitudes of the Perihelion and Nodes are increased at the rate of 50¼" each year by the Precession of the Equinoxes. The sixth column shows the angle made by the plane of the Orbit with the plane of the Ecliptic, the points of intersection being at the Nodes.

Name.	Place of Perihelion.	Annual Variation.	Place of North Node.	Annual Variation.	Inclination of Orbit.	Annual Variation.
Mercury,	♊ 15° 30' 48"	+ 5.81"	♉ 16° 50' 39"	— 7.82"	7° 0' 18"	+ 0.181"
Venus,	♌ 9° 42' 32"	— 2.68"	♊ 15° 33' 6"	— 18.71"	3° 23' 32"	+ 0.045"
Earth,	♋ 10° 16' 38"	+ 11.81"	....	....	....	....
Mars,	♎ 3° 45' 28"	+ 15.82"	♉ 18° 33' 16"	— 23.29"	1° 51' 6"	— 0.003"
Jupiter,	♐ 12° 18' 47"	+ 6.65"	♋ 9° 21' 27"	— 15.81"	1° 18' 35"	— 0.226"
Saturn,	♑ 0° 35' 23"	+ 19.37"	♋ 22° 34' 37"	— 19.42"	2° 29' 24"	— 0.155"
Uranus,	♍ 18° 30' 8"	+ 2.4"	♊ 13° 17' 0"	— 36. 0"	0° 46' 30"	+ 0.031"
Neptune,	♑ 11° 19' 28"		♌ 11° 9' 30"		1° 47' 2"	

The following table shows the diameter in miles, and the angular diameter of each body, in seconds, when at the mean distance from the Earth; the weights of each as compared with those of the Sun and Earth, and the Densities as compared with that of the Earth, and with equal bulks of water.

	Diameter in Miles.	Diameter in Seconds.	Weight Sun = 1.	Weight Earth = 1.	Density Earth = 1.	Density Water = 1.
Sun,	851736	1923.6"	1.000000	354936.	0.284	1.533
Mercury,	2960	6.7"	1/4865751	0.0729	1.392	7.518
Venus,	7566	17.1"	1/390000	0.9101	1.032	5.572
Earth,	7925.6		1/354936	1.0000	1.000	5.4
Mars,	3900	5.8"	1/2680637	0.1324	1.105	5.965
Jupiter,	88316	38.4"	1/1017.859	338.718	0.258	1.393
Saturn,	71036	17.1"	1/3512.3	101.364	0.149	0.804
Uranus,	34704	4.1"	1/24905	14.252	0.10	1.025
Neptune,	32243	2.4"	1/18780	18.98	0.335	1.807

The following are the Elements of the Moon, and of her Orbit.

Mean Distance in Radii of Earth,	59.96435	Sidereal Revolution, days,	27.321661418
Mean Distance in Miles,	237,626	Synodical Revolution,	29.530588715
Eccentricity of Orbit,	0.054844	Inclination of Orbit,	5° 8' 47.9"
Diameter in Miles,	2153	Revolution of Nodes, days,	6798.28
Angular Semidiameter,	14' 44" to 16' 46"	Revolution of Perigee,	3232.57534
Weight (Earth = 1),	0.011399	Density (Earth = 1),	0.5657
Weight of Earth and Moon (Sun being 1), =	....		1/354936

OTHER MEMBERS OF THE FAMILY.

214. PLANETOIDS.—A great number of smaller bodies have been discovered within the present century, revolving around the Sun in orbits between those of Mars and Jupiter. Four of these —named Ceres, Pallas, Juno, and Vesta — were known as early as 1807; their periods of revolution vary from 3 years 7½ months to 4 years 7½ months, and their average distances from the Sun are so nearly the same that their orbits may be compared to a number of hoops of nearly equal size, each intersecting every other in two opposite points. They received the name of Asteroids (star-like), and have since been called Planetoids (planet-like).

For thirty-eight years these four were supposed to be the only members of this singular sub-family of bodies, but in 1845 a fifth was found, and since then the number has been increased to about 100; to the end of 1866, ninety-one had received names. The orbits of all these bodies have been computed, but it is not necessary here to give directions for finding them among the fixed stars. The orbit of Vesta is relatively so small that she is sometimes much nearer the earth than any other of the larger Planetoids, and is then faintly visible as a white star-like body of the 6th magnitude. The others are, at all times, invisible to the naked eye.

The orbits of about one-third of the number form angles of more than 8 degrees with the plane of the Ecliptic, and consequently are often outside the Zodiacal limits. (See. 27). The inclination of the plane of Pallas' orbit is nearly 35 degrees.

215. The following are the elements of the principal Planetoids, for the end of the year, 1856:

Planetoid Discovered.	Ceres. January 1, 1801.	Pallas. March 28, 1802.	Juno. September 1, 1804.	Vesta. March 29, 1807.
Mean Distance,	2.765765	2.769533	2.668644	2.360559
Eccentricity,	0.079180	0.239045	0.256535	0.090164
Sidereal Revolution; days,	1,680,047	1,683,481	1,592,305	1,324,710
Perihelion,	♌ 29° 34′	♌ 2° 4′	♐ 24° 8′	♐ 10° 16′
North Node,	♊ 20° 48′	♍ 22° 38′	♍ 21° 0′	♋ 13° 24′
Inclination of Orbit,	10° 36′	34° 43′	13° 3′	7° 8′
Diameter in miles,	283	214	140	284
Apparent Magnitude,	8th	7th	8th	6th

Another sub-family of Planetoids is believed to revolve rapidly round the Sun within the orbit of Mercury, and still another band of much smaller bodies revolve in a very prolonged elliptic orbit, which, at the side nearest to the Sun, crosses the Earth's path at a point passed by her in August, and at the other side stretches far away beyond the orbit of Neptune (See Sec. 229). A second orbit, similar to the last named, is believed to be partially filled up, the Planetoidal particles performing one revolution in about 33½ years, and forming a band of at least 50,000 miles in thickness, and 1,000,000,000 miles in length, or about one-fourth the circumference of the entire orbit. This crosses the Earth's path at a point passed over by her in November. It is highly probable that thousands of other zones, or parts of zones, revolve around the Sun at different distances, and it is more than possible that our Earth has a like accompaniment. The planet Saturn is surrounded by an immense ring of matter, so dense that it has been thought to be solid, but it has appeared at times to be divided into three or more bands, and is undoubtedly composed of separate bodies, all revolving around the planet in an annular orbit.

216. SATELLITES.— Several of the Planetary bodies are themselves centers of revolution for one or more worlds. The Earth is attended by the Moon, and it is considered probable that she has another, but very small, attendant, revolving around her at about two-fifths the distance of the Moon. Neptune has one satellite, Uranus six or seven, Saturn has eight, besides a zone of Moon matter (See. 215), and Jupiter has four. These Satellites all revolve about their primary planet, or rather about their common center of gravity (Sec. 196), and accompany it in its journey around the Sun. After our own Moon the Satellites of Jupiter are of the greatest interest to us, as the instant of their apparent contact with the edge of the planet is frequently watched by the navigator for the purpose of finding the difference in time, and hence the distance (Sec. 191), between two different points on the Earth's surface.

217. The following are the elements of the four Moons of Jupiter:

No.	Sidereal Revolution.	Distance in Radii of Jupiter.	Apparent Diameter.	Diameter in Miles.
First, . . .	1d 18h 27m 33.5s	6.04853	1.015"	2436
Second, . .	3d 13h 13m 42s	9.62347	0.911"	2187
Third, . .	7d 3h 42m 33.4s	15.35024	1.488"	3573
Fourth, . .	16d 16h 32m 11.3s	26.99835	1.273"	3057

218. COMETS.— A mass of luminous matter, attended by a stream of light, is sometimes seen in the heavens. This is called a Comet, from the hair-like appearance of its train. Comets appear at very irregular intervals, with varying degrees of brilliancy, and are often seen outside the Zodiacal limits (Sec. 27) which include the orbits of all the larger planets, and of about two-thirds of the Planetoids (See. 214). For these reasons they have been regarded by some as the guerillas of the Universe, but it is known that they are subject to the laws which govern the motions of the planetary bodies, though astronomers have not been able to map out their orbits with the same precision.

The paths and periods of about 240 Comets have been computed, with greater or less accuracy, and it is estimated that several thousand of these objects come near enough to the Earth's orbit to be visible under favorable circumstances, their periods of appearance ranging from a few months upwards, to hundreds of years. The total number circulating within the bounds of the Solar System is supposed to be more than three millions.

219. Comets revolve around the Sun, in orbits of elliptic shape,— some of which are very much elongated, having one focus in, or near, the Sun, and the other many millions of miles distant. When in Perihelion (See. 177), the Comet is often a very brilliant object, and its train is many degrees in length, spreading over a large part of the firmament. The Comet is then receiving so much light from the Sun as to become luminous, and moves very rapidly (See. 201). After having passed the Perihelion, the light received diminishes, the train rapidly loses its brightness, and the angle under which it is seen from the Earth becomes smaller as the distance from the Sun increases, till at last the small portion of light reflected from the Comet is lost in the longer passage to the Earth, and the Comet is not again seen till, after having swept far out into space, it returns near the Earth's annual path.

When in that part of the orbit nearest the Aphelion, the Comet moves much more slowly than when nearest to the Sun, the velocity being (See. 201) inversely proportional to the Square of the Radius Vector; the Comet is, therefore, invisible during a great portion of the time occupied in each revolution.

220. The first Comet of whose appearance we have any knowledge (January, A.D. 66), is remarkable, also, as having been the first to be identified as a member of the Great Solar Family; this fact was ascertained by Dr. Halley soon after its re-appearance in 1862, and the Comet has, therefore, been named after him. It revolves about the Sun once in 76½ years. Its Perihelion distance is about 53,570,000 miles; the distance at Aphelion is estimated to be 3,243,000,000 miles, or 500,000,000 miles outside the orbit of Neptune. The motion is retrograde (See. 152). The following shows the results of the observations made on the three last returns to the Perihelion; the distance at that point being compared with the Earth's mean Radius Vector as Unity:

	1682.	1759.	1835.
Date of Perihelion,	Sept. 15 — 7 A. M.	March 13 — 1 A. M.	Nov. 16 — 10¼ A. M.
Place of Perihelion,	♒ 1° 36	♒ 3° 10	♒ 4° 32
Place of North Node,	♉ 21° 11	♉ 23° 50	♉ 25 10
Inclination of Orbit,	17° 45	17° 37	17° 45
Perihelion Distance,	0.5829	0.5845	0.5866

These observations show that the Longitude of the Perihelion and place of the Nodes, and the Perihelion distance, are constantly increasing. The elements of the orbit are computed from a series of observations taken before and after the Perihelion passage; these give a portion of the curve, from which the whole is ascertained (See. 199).

221. Encke's Comet revolves in a comparatively small orbit, in the short period of 3 years and 4 months. It was first seen in the year 1786, and the elements of the orbit were computed by Encke from observations on its Perihelion passage, January 27th, 1819. The Perihelion distance is about 31,000,000 miles, its place about 7° of ♏. The Aphelion distance is 370,000,000 miles.

Another Comet of short period, named after Biela — who determined its elements on the occasion of its Perihelion passage March 18th, 1826 — is chiefly remarkable as having excited fears of collision with the Earth during the present century. Its period is 2,460 days, or 6 years 9 months; the Perihelion distance is about 80,000,000 miles; the distance at Aphelion is 505,000,000 miles; its orbit very nearly touches that of the Earth in a point passed by the latter about the end of November. This Comet was partially divided in December, 1845, probably by collision with a stream of planetoids (Sec. 215).

222. The quantity of matter in a Comet is very small, as compared with the least of the planetary bodies; the fixed stars have often been seen through the most dense parts of the train, and the head, or nucleus itself, often appears to be little more than an aggregation of vapor. That a Comet has but little comparative weight, is also evident from the fact that when very near to a planet it has no appreciable power to turn that body from its normal path (Sec. 204), while more than one instance has been noted in which the Comet has been turned aside by the attraction of the planet, and forced into a new orbit.

Comets are, therefore, believed to be composed of matter in a state of great tenuity — volumes of gaseous substance, or small particles in a solid form, separated many miles from each other. In the latter case the character of the aggregation is identical with that of the Planetoid grouping (Sec. 215), and it is thought probable that two Comets, — the third of 1862, and the first of 1866 — are comparatively dense portions of the two streams already mentioned, and revolve with them in periods of 121 years 6 months, and 33 years 2¼ months.

The other portions of the Planetoidal stream, failing to reflect the light of the Sun, are necessarily much less dense than the Comet, containing even less matter than an equal bulk of our atmosphere, and averaging but a few grains in weight to the cubic mile, though occasionally a comparatively large mass may occur. Matter so very thinly spread out can exercise very little influence on a solid Planetary mass, and it is highly probable that if a Comet should come into direct "collision" with the Earth, we should feel no more shock than if a large stone fell to the ground from a height of two or three hundred feet.

223. AEROLITES. — When the Earth, in her annual journey, meets with a stream of Planetoids, those particles nearest to the Earth are drawn out of their course by her superior attraction, and fall to her surface. In passing through the atmosphere they become heated by the friction of the air, and, abstracting oxygen from it, many of them burn with great rapidity. The larger masses fall, intact, to the ground, in which they are sometimes found imbedded to the depth of several feet. The smaller particles are burned up, and their ashes fall, more slowly, in the shape of minute atoms of dust.

Few clear nights pass without the visible fall of many of these aerolites, and there is no reason to doubt that they fall as numerously during cloudy nights and in the day time. Sometimes they descend in showers, according as the particles met by the Earth are more or less numerous; the most brilliant displays occur in August and November (Sec. 215). The aerolites of August seem to spring chiefly from a point about 7 degrees South of Algol in Perseus; those of November appear to radiate from a point near Algeiba in the Sickle of Leo. We estimate the thickness of the stream of Planetoids to be 50,000 miles (Sec. 215), for the reason that the aerolites continue to fall while the Earth passes over that distance in her orbit.

This almost incessant fall of aerolites proves that, at least some, portions of what are usually called the "regions of space," are comparatively full of these floating particles, each of which is constantly in motion around some point as a center of revolution, just as the Moon moves around the Earth, and as both the Earth and Moon perform an annual journey around the Sun.

224. It has been estimated that the number of aerolitic bodies falling towards the Earth, is about eight millions daily, possessing an aggregate weight of 1,000 pounds, or an average of a little less than a grain each. This is equal to an increase of the Earth's mass to the extent of 180 tons annually. About nine-tenths of this matter is supposed to descend to the ground, or the water, in the shape of dust (Sec. 223); the remaining one-tenth part retains the solid form on reaching the surface, the individual weight of these masses varying from a single ounce to more than 17 tons. One of these bodies, weighing 700 pounds, fell at Concord, Ohio, on May day, 1860, and another of 300 pounds weight fell at Weston, Connecticut, in December, 1807.

A total of eighteen solid aerolites are known to have fallen in the United States during the first sixty years of the present century, their aggregate weight being about 1,250 tons, and their average weight $3\frac{2}{3}$ times as great as that of an equal bulk of water; their specific gravity being thus somewhat greater than that of the Moon (Sec. 196).

There is no reason to suppose that these averages will not hold good for the Planetoids falling over the whole extent of the Earth's surface.

225. The fall of an aerolite has been witnessed on several occasions, and the body has been found, in every case, to be very hot. On being subjected to chemical analysis they all prove to be composed of substances identical with those called chemical elements, the combinations of which make up the sum total of all the various forms of matter within our reach. Some of them contain a large proportion of iron, copper, nickel, tin, sulphur, phosphorus, oxygen, and the components of salt, clay, flint, lime, potash, and coal. None have yet been ascertained to contain any element not already known, though all, without exception, contain a compound of iron, nickel, and phosphorus, called Schreibersite, which has never been met with, *as a compound*, except in aerolites. Some of those which are principally composed of the least cohesive substances are broken up by the heat produced in the fall through the atmosphere, and reach the ground in fragments.

DEFINE AND EXPLAIN (the figures refer to the sections):

214. Planetoids; when discovered; number; peculiarities; rings of Planetoids; Saturn's ring. 215. Elements of Ceres, Pallas, Juno, and Vesta. 216. Satellites. 217. Moons of Jupiter. 218. Comets; appearance; movements; number known; number in the System. 219. Orbit of Comet; relative motion. 220. Halley's Comet. 221. Encke's and Biela's. 222. Matter in a Comet; consequences of a collision with the Earth. 223. Aerolites; number; displays of August and November. 224. Weight of aerolites. 225. Chemical analysis.

THE FIXED STARS.

226. The attempt to ascertain the distances and bulks of the Fixed Stars, has not hitherto been so successful as in the case of the principal members of the Solar family. We can calculate our relative distance from a planet, as Jupiter, (Sec. 203)—but a journey round the whole circumference of the Earth's orbit furnishes us with no two positions in which there is a measurable change in the apparent magnitude of a fixed star, and only gives a barely appreciable difference of place in the case of a very few stars.

Careful observations of Sirius—one of the brightest of the starry host—(Sec. 79) have enabled astronomers to detect a slight difference in his apparent declination, as seen from opposite sides of the Earth's orbit, equal to a parallax (Sec. 177) of about one second and a quarter, taking the Earth's mean distance from the Sun as a base line; this angle is less than one-seventh part of the Solar Parallax, the base line of which is the Earth's radius. The distance of Sirius from the center of the Solar System, as computed from this parallactic angle (Sec. 194) is something more than fifteen millions of millions of miles. *α* Centauri is supposed to be the nearest star to our Sun.

If we should assume that the fixed stars are all of equal bulk, it would follow that those of lesser apparent magnitude must be much farther distant. It is, however, probable that the stars differ as much in actual size and weight, as do the heavenly bodies whose bulks have been measured, but it is also probable that some of the largest are among those of small apparent magnitude, and it has been estimated that Sirius, and a few other Stars which have been found to show a small parallactic angle, are as much nearer to us than a multitude of stars which have no parallax, as is the Moon compared with Neptune.

It has been thought that some of the more distant of those stars visible to the naked eye are at least sixty times farther removed than Sirius from the Sun. Even these remote bodies do not lie near the boundaries of the Universe. The cloud-like nebulæ in Cancer, and Andromeda, and Orion, which are visible at times to the naked eye; the shining Milky Way which encircles the heavens, and thousands of other luminous groupings, have been looked at through the telescope, and seen to consist of distinct Stars, only shut out, by their remoteness, from the unaided vision.

It is probable that if we could stand on one of those most distant Stars, and look out farther away from the Sun, we should still find the view bounded only by our ability to peer into the immensity of space. We conclude that the Universe is infinite in extent; without bounds. We may not be able to grasp the full meaning of this statement, but there is no more difficulty in believing it than in trying to think of a vast space with unknown limits, and the inevitable *something beyond.*

227. The number of fixed stars discernible with the naked eye, is about three thousand. Of these 19 are of the first magnitude; 3 between the first and second; 31 of the second; 25 between the second and third; 116 of the third; and 73 between the third and fourth. On the preceding maps they are represented as being 22 of the first,

56 of the second, and 189 of the third. The numbers classed under each successive magnitude increase rapidly, and when the telescope is employed the Stars may be counted by millions, spaces being found to be thick with Stars, which were mere blanks to the naked eye. The number seen appears to be limited only by the power of the instrument used. We conclude that the Stars are infinite in number, like the extent of the space they occupy.

228. The Fixed Stars could not be visible to us at such immense distances (Sec. 226), even as luminous specks, without having great actual magnitude, and shining with their own light, like the Sun. The Planet Jupiter is 88,316 miles in diameter, and yet, at his mean distance from the Earth, he is not much brighter than Sirius, though 30,000 times nearer.

Light reflected from a body a thousand times larger than Jupiter, would be lost long before reaching us from the distance of Sirius. The light we receive from Sirius has been estimated at one part in 20,000,000,000 of that received from the Sun, but, as the intensity of light diminishes — like the attraction of gravitation — with the square of the distance, the diameter of Sirius must be twice as great, or the intensity of light from an equal area of surface must be four times as great, as that of the Sun, in order to produce this result. Hence we conclude that some of the fixed stars are at least as large as our Sun, and that, like him, they shine with their own light.

229. Many stars appear to be double; and there are few of the larger magnitudes but have a companion, in some cases so small, or so distant, as to be visible only with the aid of a good telescope. There is no doubt that in many of these instances the apparent companionship is simply an accident of position, the smaller star being many millions of miles farther away from us than its more illustrious neighbor, but nearly on the same line of vision.

There are, however, many well known cases of two, three, four, or more stars, forming a separate system. The Stars composing one of these systems are observed to approach and recede from each other, changing their relative positions with a regularity which leaves no room to doubt that they revolve around a common center of gravity, as is the case with the Earth and Moon (Sec. 196).

230. The annual variations in the Right Ascension and North Polar Distance, given in our table of Fixed Stars, are principally the equivalents of the precession in Longitude (Sec. 208), but they also include another element — an individual motion — a relative change of place, in many of the fixed stars. Most of the Stars which have been watched for a long time are found to be affected with this "proper" motion.

Reasoning from an analogy which has a wide spread basis, we find no room to doubt that every Star in the Universe is in motion around some central point or body. We conclude that motion is the condition of existence. Gravitation is the law of matter (Sec. 202), and motion is the exponent of its all-pervading power. The word "fixed," as applied to the Stars, is, therefore, to be understood in a comparative sense only, the apparent changes in the angular positions of the Stars, with regard to each other, being very small as compared with those of the planetary bodies.

231. Changes in the apparent magnitude of Stars are numerous. Mira, in Cetus, is remarkable as varying from the second magnitude to the seventh, with a period of about eleven months. Algol, in Perseus, has a shorter period of a little less than three days, during which it varies between the second and fourth magnitudes. Sheliak, in Lyra, has a periodical change in apparent brightness of about six days and a half.

From these, and numerous other cases of variation, we conclude that all the Stars rotate on their own axes, as is the case with every member of the Solar System (See. 192), and that some portions of the surface of a variable Star are less luminous than others, the brighter side being turned towards us when the magnitude appears to be greatest. It is, however, possible that the dimness may be caused, in some cases, by the passage of a planet, or belt of Planetoids, between us and the Star.

It is thought probable that each of these Stars is a Sun like our own (Sec. 228), and gives light to a family of planetary worlds, which shine with a reflected light only, and are, therefore, invisible to us at such immense distances. To the inhabitants of those worlds our Sun would appear but as a fixed star, and his large train of planetary attendants remain forever unknown.

232. Sirius is usually described as a brilliant white star (Sec. 79), but was referred to by the ancients as of a fiery red; it is now assuming a green color. Capella was also a red star formerly, afterwards yellow, and is now white (Sec. 68) but is gradually assuming a blue tinge. These, and other, variations in color warrant the inference that extensive changes are in progress. Several instances are on record of stars which suddenly appeared in the firmament, and after shining for a brief period, with rapid changes of color, have disappeared with equal suddenness, and left no traces of their existence.

The Star η, in the stern line of Argo, has recently diminished very much in apparent brilliancy, while a nebula surrounding it has correspondingly increased in brightness, and also undergone a marked change in form. The nebula is apparently growing richer at the expense of the Star. This is a very singular phenomenon; it is not only without a known parallel, but it indicates a process directly the reverse of that by which our Earth and Sun are continually receiving additions to their masses from the falling aerolites (Sec. 224).

233. The chemical analysis of the Planetoids which have fallen to the Earth's surface shows that of the sixty-three substances, called elements, of which the material of our Earth is composed, about twenty-two have been found in the aerolites, while no element has been found in their analysis, which was not previously known to the chemist (Sec. 225). This fact has been accepted as an argument in favor of the theory that all the Planets and Planetoids were formed from the same material as our Earth, though these elements may enter into different combinations, resulting in distinct forms of material existence, in the case of each body in the Solar System. Recent experiments, by which the light of the Sun has been compared with that arising from the burning of a large number of earthly substances, seem to warrant the conclusion that the Solar body is composed of the same chemical elements as those which enter into the composition of the bodies revolving around him. There are, also, good reasons (Secs. 228 and 231) for believing that the Sun and Fixed Stars are alike in character and function. Hence it is considered to be highly probable that the whole Universe of Stars and Suns, Planets, Satellites, Planetoids and Comets, was originally formed from the same elements — individualized out of a primeval chaos.

DEFINE AND EXPLAIN (the figures refer to the sections):

226. Fixed Stars; Parallax and distance of Sirius; relative distances; visible nebulæ; extent of the Universe. 227. Number of Fixed Stars. 228. Actual magnitudes; light. 229. Double Stars; relative motion. 230. Proper motion; meaning of the phrase "Fixed Star." 231. Rotations on Axis; Planetary systems. 232. Changes in color; disappearance. 233. All composed of like material. Experiments on Sunlight.

THE SUN'S MOTION.

234. The Stars in the vicinity of the constellation Hercules, are apparently farther apart now than at the beginning of the present century. Such a systematic widening could scarcely result from any other cause than a gradual lessening of the distance between us and the Stars in that part of the heavens (Sec. 199), and it is inferred from this that the Sun is moving towards that constellation at the rate of about 120,000 miles per hour, or nearly 3,000,000 miles per day, or almost twice the distance that the Earth travels in her orbit in the same time.

As every other body appears to move in a curve, it is reasonable to suppose that this motion of the Sun is in an elliptic orbit, and it has been boldly guessed that the center of this orbit is Alcyone, the brightest star in the Pleiades of Taurus, which is assumed to be 34,000,000 times as far from the Sun as the Sun is distant from the Earth. The Sun is supposed to require about 18,200,000 years for one revolution, and the plane of the orbit to be inclined 84 degrees to the plane of the Ecliptic, the two planes being nearly perpendicular to each other, and intersecting in 22° of ♉ and ♏.

235. The direction of the Sun's motion is thus nearly in the direction of the North Pole of the Ecliptic, and nearly perpendicular to the plane of the Earth's annual revolution. Owing to the great size of the Sun's Orbit, his course can vary but little from a straight line in several centuries.

The absolute motion of the Earth in space is, therefore, like the movement up a spiral staircase, and will be best understood by supposing the table (Sec. 13) to be raised through a space equal to about six times its diameter, while the ball and wire are being carried once round the edge of the table. The radius of the Earth's Orbit (91,328,000 miles) × 2 × 3.14159, gives the circumference of the orbit, which, divided by the number of hours in a year, gives 65,400 miles per hour as the Earth's mean motion. A similar computation for the Sun will give his hourly motion as 122,260 miles. But these quantities represent the two legs of a right angled triangle, and the square root of the sum of their squares represents the hypothenuse, equal to 138,700 miles, which is the absolute hourly motion of the Earth in space—being nearly twice as great as her motion in the orbit would be if the Sun were stationary.

The annual forward motion of the Sun is 11.74 times his mean distance from the Earth, and her real path may, therefore, be approximately represented by a string wound spirally round a cylinder, in such a way that the distance between the turns of the string is nearly six times the diameter of the cylinder.

If the direction of the Sun's motion were exactly perpendicular to the plane of the Ecliptic, a line parallel to it, and passing through the Earth, would always make an angle of about 28°10 with the line of the Earth's absolute motion; but this angle is subject to a slight variation, and the Earth's motion is also variable, because, owing to the slight obliqueness of the grand axis, her Radius Vector (Sec. 199) is, twice in each year, 6° out of the perpendicular to the axis.

236. The mean distance of the Moon from the center of gravity, which moves equally round the

Sun (Sec. 196), is about 235,000 miles, which is, therefore, the mean length of her Radius of Revolution. The Moon's Sidereal Period being 27.32166 days, the mean hourly motion in her orbit would be 2,250 miles, if the center of her orbit were stationary. But it is evident that the actual motion is much greater, as, during each hour, the Moon moves through a distance of 138,700 miles in space, in company with the Earth (Sec. 235). It is also apparent that her absolute motion is in a curve which differs but little from that traversed by the Earth, though relatively she revolves round a point within the Earth's mass.

Let us suppose the Earth to be represented by a marble, moving along the middle of a race course one mile in circumference, and the Moon at Full represented by a small pea placed two feet from the marble towards the outside of the track; remove the marble 35 yards along the track, placing the pea two feet in advance, on the same line, and the two will represent the positions at the Moon's third quarter; remove the marble 35 yards still farther, placing the pea two feet distant towards the inside of the track, to represent the position of the New Moon; the process continued will give the relative positions for the first quarter, and the Full Moon.

A line traced through the points successively occupied by the pea, while the marble is carried 140 yards along the track, will represent the actual path of the Moon in the heavens during one Lunation, while an orange moving more slowly in an arc half way between the track and the center of the field, will nearly represent the relative motion of the Sun with regard to the Earth and the Moon, though not with reference to the centre of his own orbit. It is apparent that the path of the Moon in space is never convex towards the Sun.

237. The relative motions of all the planetary bodies in space can not be represented by a diagram, but we can form some idea of the great movement if we conceive a number of concentric cylinders, the diameters of which are proportional to the distances of the planets from the Sun, and suppose that the rate of motion parallel to the axis of the cylinders is the same in each case (subject to the irregularity due to the fact that the planes of their orbits do not exactly coincide with each other, or with the plane of the Ecliptic), while the angular motions around the common axis of the cylinders are unequal, being proportional to the times of revolution of the bodies around the Sun (Sec. 213). If with this we can also grasp the idea of satellites revolving around some of the planets, and comets and streams of Planetoids revolving in more eliptic orbits about the ever progressing point in the axial line, we shall be able to comprehend, though faintly, the motion of the Solar system as a whole.

DEFINE AND EXPLAIN (the figures refer to the sections):

234. Absolute motion; Orbit of the Sun. 235. Motion of the Earth in space; illustration. 236. Moon's actual motion. 237. Relative movements of members of the Solar System.

INDEX.

[The numbers refer to the *pages*, not to the sections.]

A.

Achernar (in Eridanus); 21, 46.
Acubens; 26, 29, 48.
Adhara (in Canis Major); 26, 47.
Aerolites; 94.
Agena; 42, 44, 49.
Albireo (in Cygnus); 38, 43, 52.
Alchiba; 31, 48.
Alcor; 29.
Alcyone (in Pleiades); 22, 46.
Aldebaran (in Hyades); 22, 46, 57, 59.
Alderamin; 15, 43, 52.
Alhafara (in Sickle); 29, 48.
Alga; 31, 51.
Algedi; 36.
Algeiba; 29, 48.
Algenib; 17, 21, 38, 43, 45.
Algol; 18, 22, 46, 97.
Algorab (in Corvus); 31, 49.
Alhena; 25, 47.
Alioth; 13, 28, 49.
Alkaid; 13, 28, 39, 49.
Alkaturgos; 39, 50.
Alkes; 31, 48.
Alnisach; 17, 18, 46.
Alnilam (Belt Orion); 23, 47.
Alnitak   do.   ; 23, 47.
Alphard (Cor. Hydrae); 26, 31, 41, 48.
Alphecca; 39, 40, 50.
Alpheratz; 17, 18, 37, 38, 45.
Alphirk; 15, 52.
Alruccabah (Pole Star); 13, 45, 53.
Alshein (in Aquila); 35, 52.
Altair; 35, 52.
Altar (see Ara).
Aludra; 26, 41, 47.
Alwaid; 15, 51.
American Goose; 53.
Andromeda; 17, 18.
Angular Measures; 9, 10, 13, 17, 73.
Annual Motion; 6, 7, 8.
   " Variation; 45 to 53, 90.
Annular Eclipse; 75.
Anser; 42, 43, 52.
Antarctic Circle; 42.
Antares; 32, 33, 50.
Antinous et Aquila; 35, 37, 42.
Aphelion; 73.
Apogee; 73.
Aquarius; 9, 37, 38.
Aquila; 35, 36, 37, 42.
Ara; 32, 33, 34, 36, 42, 44.

Archer (see Sagittarius).
Arcturus; 30, 39, 49.
Argo; 27, 41, 42, 44.
Aries; 9, 17, 19, 20, 22.
Arista; 30, 39, 49.
Arneb; 26, 47.
Arrow (see Sagitta).
Ascensional Difference; 55, 56, 59.
Ascension (see Right Ascension).
Aselli (in Cancer); 26, 48.
Asterion; 28, 39.
Attraction of Gravitation; 84.
Attraction — Tides; 87.
Auriga; 23, 25, 42.
Azha (in Eridanus); 21, 46.

B.

Balance (see Libra).
Baton Kaitos; 21, 46.
Bear Driver (see Bootes).
Bears and the Pole; 16.
Beehive Nebula (see Praesepe).
Bellatrix (in Orion); 23, 46.
Benetnasch (see Alkaid).
Betelgueuse; 23, 25, 47.
Biela's Comet; 93.
Bootes; 28, 30, 39.
Bull (see Taurus).
Bull's Eye (see Aldebaran).
   " North Horn (see El Nath).

C.

Camelopardalus; 18, 27.
Cancer; 9, 26.
Canes Venatici; 28, 39.
Canis Major; 25, 26, 41.
   "   Minor; 26.
Canopus; 24, 41, 47.
Capella; 23, 46.
Capricornus; 9, 36, 37.
Caput (Head) Medusae; 22.
Cassiopeia; 14, 17, 18.
Castor; 25, 47.
Cebelrai; 34, 51.
Centaurus; 31, 32, 33, 42, 44.
Cepheus; 15, 18, 43.
Cerberus; 39.
Ceres; 91.
Cetus; 19, 20, 38.

Chameleon; 42.
Chair; 14.
Chaph; 14, 45.
Chara; 28, 39.
Charles' Oak (see Robur Caroli).
Chronology; 76.
Circinus; 34, 42.
Clypei Sobieski; 35.
Collision with a Comet; 94.
Columba Noachi; 27.
Cœlure; 13, 14, 15.
Coma Berenices; 31.
Comets; 92.
Compasses (see Circinus).
Conjunction, Inferior; 63.
   "   Superior; 63, 65.
Constellations; 10.
Cor Caroli; 28, 30, 39, 49.
  " Leonis; 28, 48.
Corona Australis; 36.
   "   Borealis; 34, 39.
Cor Scorpii; 33, 50.
Corvus; 31.
Crab (see Cancer).
Crane (see Grus).
Crater; 29, 31.
Crow (see Corvus).
Crux; 42, 44.
Culmination; 11, 54, 60.
Cup (see Crater).
Cursa; 24, 46.
Cycles; 76, 77, 78.
Cygnus; 15, 16, 38, 43.

D.

Dabih; 36, 52.
Day, Length of; 55, 59.
   " Sidereal; 6.
Declination; 8, 53, 54, 57, 58.
Delphinus; 35, 37.
Deneb; 16, 43, 52.
Deneb El Okab (in Aquila); 51.
Denebola; 29, 48.
Density of Comets; 94.
   "   " Earth; 80.
   "   " Moon; 81.
   "   " Planetoids; 94.
   "   " Sun and Planets; 90.
Deshabeh; 36, 52.
Diagram, Diurnal Motion; 55.
   "   Earth's Motion; 7.

INDEX.

Diagram, Earth's Orbit; 83.
" Horizon Circles; 61.
" Jupiter's Motion; 64.
" Lunar Motion; 72.
" Mercury's Orbit; 66.
" Moon's Distance; 80.
" Nutation and Precession; 86.
" Transit of Venus; 82.
" Venus' Orbit; 65.
Diameter of Earth; 5, 79.
" " Moon; 80, 90.
" " Planets; 85, 90.
" " Sun; 82, 90.
Diamond of Virgo; 30.
Dionysian Period; 78.
Diphda; 24, 45.
Dipper; 13, 28, 29.
Distance of Moon; 80.
" " Sun; 81, 89.
" " Venus; 84.
" North Polar; 8, 45, 53.
Diurnal Arc; 55, 59.
" Motion; 5, 7, 54, 55.
Dog, Greater (see Canis Major).
" Lesser (see Canis Minor).
Dog Star (Sirius); 26.
Dolphin (see Delphinus).
Dominical Letter; 77.
Doradus; 24, 42, 46.
Dorsa Leonis; 29.
Dove (see Columba).
Draco; 14, 15, 39.
Dragon (see Draco).
Dubhe; 13, 28, 48.

E.

Eagle (see Aquila).
Earth's Annual Motion; 6, 7.
" Axis; 6, 7.
" Diurnal Motion; 5, 54, 55.
" Motion in Orbit; 84.
" Motion in Space; 99.
" Size; 79.
" Weight; 80.
Easter; 78.
Eclipses; 72.
Ecliptic; 7, 9.
Elongation; 65, 67.
El Nath; 23, 46.
El Phackra; 48.
El Rischa; 18, 19, 46.
Encke's Comet; 93.
Enif; 38, 52.
Epact; 78.
Equal Areas in Equal Times; 83.
Equation of Time; 8, 12.
Equator; 6
Equinoctial; 6, 54.
Equinoctial Colure; 13, 14, 17, 30, 44.
Equinoxes; 7, 8, 9, 16.
Equuleus; 38.

Eridanus; 21, 22.
Er Rai; 15, 53.
Er Rakis; 15.
Etanin; 15, 51.
Evening Star; 65.

F.

Falling to the Sun; 84.
Fishes (see Pisces).
Fish, Southern (see Piscis Australis).
" Flying (see Piscis Volans).
Fixed Stars; 5, 45 to 53.
" " Changes; 97, 98.
" " Distances; 96.
" " Light; 97.
" " Number; 97.
" " Parallax; 96.
" " Proper Motion; 97.
Fly (see Musca).
Flying Horse (see Pegasus).
Fomalhaut; 38, 53.
Fox and Goose (see Vulpecula et Anser).

G.

Gemini; 9, 23, 25.
Geocentric Position; 65.
Gianser (in Draco); 48.
Gienah; 38, 43, 52.
Glenner (see Virgo).
Goat (see Capricornus).
Golden Number; 78.
Gomeisa; 26, 47.
Goose (see or Tucana); 53.
Graffias; 33, 50.
Great Bear (see Ursa Major).
Greek Alphabet; 11.
Granuium; 15, 51.
Grus; 36, 37.

H.

Halley's Comet; 93.
Hamal; 17, 19, 46.
Hare (see Lepus).
Harp (see Lyra).
Heliocentric Position; 65.
Hercules; 39, 99.
Homan; 38, 53.
Horizon; 8, 60, 61.
Hunter (see Orion).
Hyades; 22, 23.
Hydra; 28, 29, 31, 33.
Hydrus; 24, 42.

I.

Indus et Pavo; 36.
Izar; 39, 49.

J.

Julian Period; 78.
Juno; 91.
Jupiter; 62.
" Elements; 89, 90.
" Longitudes; 63.
" Moons; 92.

K.

Kaus Australis; 51.
Kochab; 14, 49.
Korneforos; 39, 50.

L.

Lacerta; 18, 43.
Lady in Chair (see Cassiopeia).
Latitude; 8, 58.
" of Earth; 87.
" " Moon; 72.
" " Stars; 45 to 53.
Law of Attraction; 84.
" " Motion in the Orbit; 83.
Leap Year; 78.
Leo; 9, 26, 28, 29.
" Minor; 29.
Length of Day; 8, 54, 55, 57.
Lepus; 25, 26.
Lesath; 33, 51.
Libra; 9, 32.
Lion (see Leo).
Lesser Bear (see Ursa Minor).
Lizard (see Lacerta).
Longitude; 8, 58.
" of Jupiter; 63.
" " Mars; 68.
" " Mercury; 67.
" " Neptune; 70.
" " Saturn; 69.
" " Stars; 45 to 53.
" " Sun; 11, 12.
" " Uranus; 70.
" " Venus; 66.
Lunar Cycle; 76.
Lunation; 71.
Lupus; 32, 34, 44.
Lynx; 27, 29.
Lyra; 40.

M.

Magnitudes of Planets; 90.
" " Stars; 10, 97.
Map I.; 14.
" II.; 18.
" III.; 21.
" IV.; 23.
" V.; 26.

INDEX.

Map VI. ; 29.
" VII. ; 30.
" VIII. ; 33.
" IX. ; 35.
" X. ; 37.
" XI. ; 39.
" XII. ; 41.
" XIII. ; 43.
" XIV. ; 60.
Mars ; 64, 89, 90.
Markab ; 17, 53.
Matar ; 38, 53.
Matter, the same in all ; 95, 98.
Measure of Angles ; 9, 10, 13, 17, 73.
" " Attraction ; 84.
" " Distance ; 73, 79.
Measures of Time ; 70.
Measure of Weight ; 80.
Mebsuta ; 25, 47.
Megrez ; 13, 17, 28, 49.
Meukalinan ; 23, 25, 47.
Menkar ; 20, 21, 46.
Merak ; 13, 28, 48.
Mercury ; 67, 89, 90.
Meridian ; 6, 60.
Mesarthim ; 19.
Meteors ; 91.
Metonic Cycle ; 78.
Microscopium ; 36, 37.
Milky Way ; 10, 14, 15, 18, 22, 24, 27, 31, 36, 38, 40, 42, 43, 44.
Mintaka (Belt Orion) ; 23, 47.
Mira ; 17, 18, 20, 21, 46, 97.
Mirach ; 17, 45.
Mirfak ; 17, 18, 22, 46, 97.
Mirzam ; 26, 47.
Mizar ; 13, 28, 49.
Monoceros ; 26, 27.
Mons Mensa ; 42.
Moon ; 71.
Moon's Age ; 78. [90.
" Distance, Size and Weight ; 80.
" Real Path ; 100.
Moons of Jupiter ; 93.
Mouth ; 77.
Morning Star ; 65.
Motion in Orbit ; 62, 83.
" of Stars ; 5 to 11, 13, 16, 97.
" " Systems ; 97, 99.
Muliphen ; 26.
Muphrid ; 39, 49.
Musca ; 19, 32, 42.
Mutual Attraction ; 85.

N.

Naos ; 41, 47.
Nebulæ ; 10, 18, 23, 26, 47, 96.
Nekkar ; 16, 39, 50.
Neptune ; 70, 89, 90.
Net (see Reticulum).
Nihal ; 26.

Nodes of Moon ; 71, 86.
Norma Euclidis ; 34.
North Asellus ; 26, 48.
North Polar Distance ; 8, 45 to 53.
Nutation ; 85.

O.

Octans ; 42.
Oculus Pavonis ; 36, 52.
Ophiucus ; 32, 33, 35, 39.
Opposition ; 63.
Orion ; 23, 25.

P.

Painter's Easel (see Pictoris).
Pallas ; 91.
Parallax ; 73.
Pavo ; 36, 42.
Peacock (see Pavo).
Pegasus ; 17, 18, 20, 38, 43.
Penumbra ; 74.
Perigee ; 73.
Perihelion ; 73.
Perseus ; 18, 22.
Phact ; 27, 47.
Pheeda ; 13, 28, 48.
Phœnix ; 21.
Phurid (in Canis Major) ; 47.
Pictoris ; 42.
Pisces ; 9, 18, 20, 38.
Piscis Australis ; 38.
" Volans ; 42.
Planetoids ; 91.
Planets ; 62, 89, 90, 100.
Pleiades ; 22, 46 (Alcyone).
Pointers ; 13.
Points of Compass ; 11.
Poles of the Heavens ; 6, 13.
Pole Star ; 5, 13, 14, 45.
Pollux ; 25, 47.
Precession of Equinoxes ; 9, 16, 85, 87.
Prime Vertical ; 60.
Procyon ; 26, 47.
Propus ; 25.

R.

Radius Vector ; 83.
Ram (see Aries).
Rasalague ; 34, 40, 51.
Rasal Asad (in Leo) ; 48.
Ras Algethi ; 34, 40, 51.
Rastaban (see Alwaid).
Ratio of Distance ; 84, 89, 90.
Refraction ; 54, 55, 57.
Regulus ; 28, 48.
Reticulum Rhomboidalis ; 24.
Retrogradation ; 62.

Rigel ; 23, 24, 46.
Right Ascension ; 8, 10, 45 to 53, 57, 58.
" " of Stars ; 45 to 53.
" " Sun ; 11, 12, 37.
Rising ; 54, 59, 60, 61.
River (see Eridanus).
Robur Caroli ; 41.
Roman Indiction ; 78.
Rotanen ; 35, 52.
Ruchbah ; 45.
Ruchbah Ur Ramih ; 52.

S.

Sadalmelik ; 37, 53.
Sadalsund ; 37, 52.
Sadr ; 43, 52.
Saiph ; 23, 47.
Sagitta ; 43, 52.
Sagittarius ; 9, 35.
Satellites ; 92.
Saturn ; 69, 89, 90.
Scales (see Libra).
Scheat Aquarii ; 37, 53.
" Pegasi ; 17, 53.
Schedir ; 14, 45.
Scorpio ; 9, 32, 33, 35.
Seasons ; 8.
Secunda Girdi ; 36, 52.
Seginus ; 49.
Semi-Diameter ; 73, 79.
Semi-Diurnal Arc ; 55, 57, 59.
Serpens ; 33, 35, 39.
Serpentarius (see Ophiucus).
Serpent Bearer (see Ophiucus).
Setting ; 54, 59, 60, 61.
Shape of the Earth ; 79.
Sheliak ; 40, 51.
Sheratan ; 19, 46.
Ship (see Argo).
Sickle of Leo ; 29.
Sidereal Day ; 6.
" Revolution ; 63, 89.
" Time ; 10, 11.
Signs of Zodiac ; 9, 10.
Sirius ; 26, 47.
Snake (see Hydra).
Solar Cycle ; 77.
" Day ; 6.
" System ; 62 to 95.
Solar Motion in Space ; 99.
Solstitial Colure ; 15, 35, 40.
South Asellus ; 26, 48.
South'n Crown (see Corona Australis).
Southern Fly (see Musca).
Southing ; 11, 54, 60.
Spica Virginis ; 30, 49.
Square of Pegasus ; 17, 18, 20, 38.
Squares of Times proportional to Cubes
 of Distances ; 85.
Sulaphat ; 40, 51.
Sunday ; 77.

INDEX.

Sun's Density ; 82, 90.
" Diameter ; 57, 82, 90.
" Distance ; 81, 89.
" Light ; 98.
" Orbit ; 99.
" Semi-Diameter ; 73, 82, 90.
" Parallax ; 73, 81.
" Weight ; 82, 90.
Svalocin ; 33, 52.
Swan (see Cygnus).
Synodical Revolution ; 63.

T.

Table Mountain (see Mons Mensa).
Table—Ascensional Differences ; 56.
" Declinations ; 58. [90.
" Elements of Solar System ; 89,
" Equation of Time ; 12.
" Fixed Stars ; 45 to 53.
" Jupiter's Places ; 64.
" " Satellites ; 92.
" Mars' " 68.
" Mercury's " 67.
" Moon's Elements ; 00.
" Neptune's Places ; 70.
" Planetoids ; 91.
" Refraction ; 57.
" Right Ascension ; 58.
" Saturn's Places ; 69.
" Sun's " 12.
" Uranus' " 70.
" Venus' " 66.
Talita ; 28, 48.
Tarazed (in Aquila) ; 52.

Taurus ; 9, 22.
Taurus Poniatowski ; 32, 34, 35.
Tejat ; 23, 47.
Telescopium ; 23, 25, 32, 33, 36.
Theemin ; 21, 46.
Thuban ; 13, 49.
Tides ; 87.
Toucan ; 33.
Transit of Venus ; 66, 81.
Triangulum ; 18, 19.
Triangulum Australis ; 33, 42, 44.
Tureis ; 27, 41, 47.

U.

Umbra ; 74.
Unicorn (see Monoceros).
Ungula ; 42, 44, 49.
Unukalhay ; 33, 50.
Uranus ; 69, 89, 90.
Urkab Ur Ramih ; 52.
Urn of Aquarius ; 37.
Ursa Major ; 13, 28.
" Minor ; 14.

V.

Vega ; 40, 51.
Venus ; 65, 89, 90.
" Distance ; 91.
" Transit of ; 66, 81.
Vernal Equinox ; 8, 13, 14, 17.
Vesta ; 91.
Virgin (see Virgo).

Virgo ; 9, 29, 30.
Vindemiatrix ; 30, 49.
Vulpecula et Anser ; 43.

W.

Water Bearer (see Aquarius).
Water Snake (see Hydrus).
Wagoner (see Auriga).
Wasat ; 23, 47.
Week ; 77.
Weight of Earth ; 80, 90.
" " Moon and Sun ; 80, 82, 90.
" " Planets ; 90.
Wesen ; 26, 47.
Whale (see Cetus).
Winged Horse (see Pegasus).
Wolf (see Lupus).

Y.

Yard Stick (see Orion).
Yed ; 33, 50.

Z.

Zaurak ; 21, 46.
Zarijava ; 29, 30, 48.
Zenith ; 11, 60.
Zodiac ; 9, 10, 20.
Zozma ; 29, 48.
Zubenelg ; 32, 50.
Zubenesch ; 32, 49.

www.ingramcontent.com/pod-product-compliance
Lightning Source LLC
Chambersburg PA
CBHW020106170426
43199CB00009B/412